Soldiers
in
Politics

Soldiers
in
Politics

Edited by

STEFFEN W. SCHMIDT
Iowa State Univ.

GERALD A. DORFMAN
Iowa State Univ.

Geron-X, Inc.   Los Altos

this book is lovingly dedicated to

Lisel Suspiero de Schmidt

Walter Schmidt

Adeline S. Dorfman

Ralph I. Dorfman

Margaret C. Dorfman

CONTRIBUTORS

Gerald A. Dorfman is Assistant Professor of Political Science, Iowa State University. He holds the Ph.D. from Columbia University. His general research interests include British and western European politics and government. He is author of *Wage Politics in Britain, 1945–1967* and both publisher and co-founder of the journals, *Politics and Society* and *Political Methodology*.

Steffen W. Schmidt is Assistant Professor, Department of Political Science, Iowa State University. He holds the Ph.D. from Columbia University, New York City. His scholarly interests are Latin American Politics, the study of Political clientelism and the political role of women. He has published in numerous scholarly journals including *Comparative Politics,* the *Journal of Latin American Studies* (Cambridge, England) and the *Journal of Inter-American Studies and World Affairs.*

Barry Ames received his Ph.D. at Stanford, taught at the University of New Mexico and currently teaches at Washington University, St. Louis. Past research concerns Mexican students, Mexican elections, and the study of bureaucracies. His current interests focus on public expenditures in Latin America between 1945 and 1971.

James Brown is Assistant Professor of Political Science at Southern Methodist University. He received his BA from Texas Christian University and his MA and Ph.D. from the State University of New York at Buffalo. He is an Associate Editor of the *Social Science Research Journal,* Athens, Greece. In addition to several articles, he will soon publish *The Military and The Politics of Greece.*

John Cady is Distinguished Professor Emeritus of History at Ohio University. Among his many works are *Roots of French Imperialism in Eastern Asia* and *History of Modern Burma.* His most recent work is *United States and Burma.*

William Eckhardt is a clinical and social psychologist working at the Canadian Peace Research Institute, Oakville, Ontario, since 1967. Prior to peace research, he worked in child guidance and mental health clinics, mental hospitals, and colleges in the United States since 1954. His recently published book *Compassion* is an attempt to establish a science of value on the basis of empirical research.

Luigi R. Einaudi has been a member of the Social Science Department at The Rand Corporation since 1962. He received his A.B. and Ph.D. degrees from Harvard and has taught at Harvard, Wesleyan, and most recently at UCLA. He is the editor of *Beyond Cuba: Latin America Takes Charge of Its Future*, Crane, Russak & Company, New York, 1973.

Edward E. Feit is Professor of Political Science at the University of Massachusetts at Amherst. His most recent book *The Armed Bureaucrats: Military-Administrative Regimes and Political Development* has just been published. He is the author of three other books and many articles in scholarly journals.

Irving Louis Horowitz is Professor of Sociology and Political Science at Rutgers, The State University. He is author of a number of works in political sociology, including *The War Game: Studies of the New Civilian Militarists;* and *Three Worlds of Development: Theory and Practice of International Stratification.* Professor Horowitz is Editor-in-Chief of Society magazine (formerly *trans*action), the leading multidisciplinary periodical in American social science.

Paul Sunik Kim, a native of Korea, is Associate Professor of Political Science at Gannon College, Erie Pennsylvania. After earning his Ph.D. from New York University, 1964, he taught at Upsala College and Susquehanna University.

David M. Krieger is at the Center for the Study of Democratic Institutions. He is the author of articles in the *Journal of Peace Research* and the *Journal of Contemporary Revolutions.*

Rene Lemarchand is Professor of Political Science at the University of Florida (Gainesville, Fla.). His publications include *Political Awakening in the Congo* (Berkeley & Los Angeles, University of California Press, 1964) and *Rwanda and Burundi* (Pall Mall Press, London: 1970). He is currently working on a book-length study of dependency relationships in Tropical Africa.

Byron T. Mook is at the department of Political Science, Duke University and Research Fellow, The Institute of Development Studies, The University of Sussex. Has done research in South Asia on bureaucratic values and behavior. Research in progress includes work on educational planning in India and Sri Lanka.

Charles C. Moskos, Jr. is Professor of Sociology at Northwestern University. In addition to numerous articles on military sociology, he is author of

The American Enlisted Man (1970), and editor of *Public Opinion and the Military Establishment* (1971).

Martin Needler is Director of the Division of Inter-American Affairs and Professor of Political Science at the University of New Mexico. He is the author of various books on Latin American politics, including *Latin American Politics in Perspective, Political Development in Latin America, Politics and Society in Mexico,* and *The United States and the Latin American Revolution.*

Adam Yarmolinsky is the Ralph Waldo Emerson University Professor at the University of Massachusetts. He served in the Pentagon and in the White House during the Kennedy and Johnson Administrations. He is the author of numerous articles and books, including *The Military Establishment* and *Recognition of Excellence.*

Claude E. Welch, Jr., Professor of Political Science at the State University of New York at Buffalo, co-authored *Soldier and State in Africa* and *Military Role and Rule,* and edited *Revolution and Political Change* and *Political Modernization.* He is currently studying peasant rebellions in developing countries.

CONTENTS

As the countries of Africa, Asia, the Middle East and Latin America struggle to become modern nations, traditional arrangements, patterns of resource
allocation, and traditional values increasingly come under strain. Modernization is painful; the fruits of change are not immediately available. Conflicts,
therefore, often erupt between different regions, classes, tribes, linguistic
groups, religious sects and even between countries, as the shifts in ideology and
in the legitimacy of who shall rule make themselves felt. Armed groups, by
their control of violence and their organizational coherence in an often
chaotic situation, are one of the key elements in these struggles.

Several kinds of armed groups contend for control of political systems
throughout the world. The most widely publicized are urban guerillas, rural
insurgency groups, and anti-colonial, nationalist movements. Yet, of equal
significance, but less well understood are the "official" armed men maintained
by every country. This volume examines the role of this latter group—the
military.

Soldiers in Politics is a collection of the most important, recent work on
the role of professional soldiers in political life everywhere. It focuses, generally, on three major questions: how does the military become involved in
politics; how does the military perform when it is in politics; and finally, how
does the military get out of politics. In turn, these general questions raise a
myriad of subordinate questions which provide the focus for the articles contained in this book. First, what are the quality and quantity of linkages between the military and the larger society in which it operates? Specifically,
how does the military deal with other functionally oriented groups which
operate the political system? How, for example, does the military relate to
existing political parties or factions? When does it employ force and what
resources other than force does the military possess, and how does it use
those non-military resources? What groups become allies to military regimes
and for what reasons—for instance, do regional, class, tribal, racial and other
factors play an important role in civil-military relations?

Other questions deal with the internal ethos and structure as well as recruiting patterns of the military itself. What is "professionalization" and does
it lead to more or less military involvement in politics? How are soldiers
(especially officers) recruited into the services? Do the regional, educational,
class and other backgrounds of officers make a difference in their political
behavior as soldiers? What sorts of variables account for intra and interservice rivalries? Are there measurable changes in the orientation of armies
over time to such issues as social justice, national self-determination and

international alliances? Assuming that the military has taken over a government what types of officers are most likely to urge a return to civilian rule? Conversely what elements within the services are most suspicious of civilian rule? Why do some military establishments promote revolutions of one type or another while others are status-quo oriented? Does the type of training officers receive have any impact on whether or not they will be politicized? If so, what types of training has what sorts of effects?

Still, a third very crucial aspect of military politicization involves linkages between domestic issues and international problems. For instance, do defeated or victorious armies get more involved in politics? Is it true that the military may intervene in politics unless it has "military" functions to perform? Put another way, if armies don't fight wars will they turn to politics as an outlet for their energy? Are military alliances with other powers a stimulus for intervention? Do foreign military missions get involved in domestic politics? Is military assistance given by developed nations to third world countries a major unsettling factor in the domestic politics of these nations? Indeed, do international tensions provide the military in *developed* countries with the opportunity to become politically influential in their own societies? Are the armies in Socialist countries as much involved in politics as those in countries linked to other ideological systems and international "fraternal" networks?

No single study herein answers all of these questions, nor does the volume as a whole provide final answers. Despite recent intense efforts, scholars find it difficult to be definitive about the military in politics. Talk of a coherent theory is premature; most generalizations are still not complex enough to account for the variables involved—over time, across national and continental boundaries or within various domestic and international contexts. Substantial additional research is still necessary. *Soldiers in Politics* thus provides a forum for various scholars to present the fruits of their research, to provide ideas upon which further work can be developed.

We have chosen to include each article because we feel it is substantively interesting; in some cases, the articles are impressionistic sketches with important heuristic value. As a whole, the articles are methodologically, ideologically and stylistically diverse. Most of them deal with five areas of the world, including the United States. Almost all of the articles are being published for the first time, and some were written especially for this volume.

Finally, we think that this book will appeal to a broad audience. We have tried, especially, to include articles that would be useful in introductory and advanced undergraduate comparative politics courses. We have also included papers and articles which report new and significant research findings of importance to professional scholars working on civil-military relations. We also hope that the cursorily-interested layman will benefit from a reading of *Soldiers in Politics*.

PART I

Soldiers
in Politics

MILITARIZATION, MODERNIZATION AND MOBILIZATION[1]

*Irving Louis Horowitz**

Military Power and State Power

Military power exists in its relation to state power. It is geared to defend a well-defined geographic terrain and a certain body of people having a common set of economic, psychological and linguistic elements within this terrain. The main function of the armed forces at the outset of nation-building is to preserve and make visible national sovereignty. They "defend" and "project" the national entity into the international arena. And once actual sovereignty is obtained, they acquire a critical role with respect to the internal affairs of the state. The military revolutionists may come to power against colonialists, but they maintain their power against internal threats, especially by preventing political infighting among the ruling groups (Meister, 1968, p. 261).

The relationship between government and the military is no less intimate than the connection between the control of power and the control of violence at the more general level. But whatever the main role of the armed forces at any given moment, whether the maintenance of the state against external enemies or against internal terrorists, the military leadership alone is assigned the right to use physical violence. Such a right does not, however, extend to legal authorization. The ambiguity of civil-military relations often resides in

* Professor of Sociology and Political Sociology, Rutgers University.
1. This article will appear as a chapter in a forthcoming book *Foundations of Political Sociology* by Irving Louis Horowitz. Harper and Row. New York 1972.

the separation of the permission to use violence (a legal-social function) from the actual conduct of violence (a military function). Such exercise of violence in the absence of complete legitimation also remains a constant source of tension between the military and the civil apparatus. The military, when functioning properly, quickly acquires a sense of the nation and becomes sensitized toward vested interests and factional enclaves that break down the egalitarian consensus of the revolutionary period. Of course, there are many cases in which the military itself becomes a vested interest and serves to unbalance equitable arrangements that may have been made during the early revolutionary period. Furthermore, when various factions of the military cleave and adopt different ideologies toward national goals, a period of protracted civil strife usually follows. This was certainly a prime factor in both the Congolese and Nigerian civil wars. While there is no automatic rule which insures the military maintenance of a postrevolutionary equilibrium, the military will assume such a role when no other social factor does so.

One way the military is able to serve national interests with relative equality is related to its methods of recruitment. The military gains membership from disparate groups and classes within the society. By minimizing the class base of membership through a heterogenous recruitment policy, the armed forces can function as an important socializing agency (La Palombara, 1963, pp. 31–32). This transclass role can be performed despite the relatively pronounced class base of military leadership. In Middle Eastern states and nations such as Egypt and Iran, where the class formations remain relatively diffuse and weak, the corresponding importance of the army as a national and even a class-welding agency becomes manifest (Halpern, 1963; Kirk, 1963).

Unlike the experience in either the United States or the Soviet Union, the armed forces in the newly emergent nations are not absorbed into the civilian society, but instead become partners with the society. One of the hallmarks of George Washington's Administration was its subjection of the military to political control (Lipset, 1963, pp. 16–45; Chambers, 1963, pp. 21–27). After the Russian Revolution, despite urgings of the permanent revolutionists, the military was placed under the rule of the political elites (Werner, 1939, p. 36). In the United States, civilian control had democratic consequences and in the Soviet Union autocratic consequences, but in neither case did it make much difference in terms of the functional efficiency of the armed forces. In both the United States and the Soviet Union the military served as a professional source of political strength and orientations toward economic development rather than as a ruling directorate. The same cannot be said for most Third World nations, where, as a matter of fact, the political functions are oftentimes militarized from the onset of independence. Due to this early identification with the national cause, the military is transformed from a symbolic badge of sovereignty into a decisive partner in the composition of the state.

The function of the military establishment as the mark of sovereignty is well exemplified by post-liberation India. Given its strong traditional bias against force and violence, India represents a good test case. Under the reign of Nehru, the Gandhian approach to pacifism was severely modified in the name of expediency. According to Gandhi, the key to real victory is the doctrine of *Satyagraha*—the force that is born of truth and love, rather than error and hatred. Nonmilitary, nonviolent social action was pitted against all enemies (Bondurant, 1958; Morris-Jones, 1960; Horowitz, 1957, pp. 89–106; Naess, 1965). But the actual conduct of foreign policy after the British left compelled a quick and uneven modification of this policy. With reference to the early stages of the Kashmir dispute, Nehru (1957, p. 357) considered it his "misfortune that we even have to keep an army, a navy, and an air force. In the world one is compelled to take these precautions." While the rhetoric remains pacifist, the actual chore of strengthening the military was well under way no more than two years after independence. And, by the end of the 1950s, Nehru (1957, p. 211) had abandoned even the pacifist rhetoric. He noted that "none of us would dare, in the present state of the world, to do away with the instruments of organized violence. We keep armies both to defend ourselves against aggression from without and to meet trouble from within." India responded to the increased military determinism of all worldly situations. It was not so much that Gandhism was a dead letter in India by Independence Day, but rather that nationalism proved to be more compelling than ideology in shaping Indian society. The bitter controversy over Pakistani independence in the 1950s and the equally dangerous rift with China in the 1960s moved India into a much more conventional statist posture than it had in its origins. But at the same time, it became more nearly a national power and lost a good deal of its claim to be an international moral authority.

The increase in military spending in India has kept pace with the general militarization of the Third World. From 1960 to 1964 alone, the increase in military allocations was nearly threefold (Ministry of Information and Broadcasting, 1962, pp. 72–74; 1963, pp. 21, 64, 180). India's defense appropriations went from 6 percent of the gross national product (GNP) in 1961 to 17 percent in 1964, and enrollment in the national military cadet corps, India's "West Point," rose from 150,000 in 1958 to 300,000 in 1964. Slowly also, the military mix of army, navy, and air force grew steadily at the expense of the other branches. And with the increasing military hardware sent in by both the Soviet Union and the United States, India was transformed from a Third World leader to a buffer zone.

The military buildup is continuing. The rise of Communist opposition to Congress Party rule in Kerala and elsewhere and the pressures from China and Pakistan only partly explain this increase. For, as in the other Third World nations, national greatness is becoming ever more linked to military grandeur. And no Third World country, even one conscientiously dedicated to pacifism, has been able to withstand this formula of a civil-military partnership; what can be controlled is the degree of military involvement (as in some parts of Africa) but no longer the fact of involvement.

The Politicized Military

The capacity of the military in Third World nations to help establish a political system depends upon three main factors: its control of the instruments of violence, its ethos of public service and national identity instead of private interest and class identity, and its representation as an articulate and expert group (Janowitz, 1964, pp. 27–28). In the Third World the military alone combines these factors, which may be generalized as technical skills combined with an ethic of national purpose.

These factors also help to explain why the elite of the armed forces no longer confine their allegiance to traditional upper classes. The rise of the middle class, as well as of the working class, in many portions of the Third World has forced the military to become more nearly representative of the nation as a whole than its own vested interests or natural proclivities might have indicated. As long as the military maintained such alignments with the aristocracy and with religious groups, it was difficult, if not impossible, for the nation to become developmentally oriented or for the military to perform in terms of public service or its technical or professional skills. This shift has been most pronounced in those nations where the military underwent a transformation from within, especially in Asian and Sub-Sahara Africa where the inherited military establishment had to break with its oligarchical notion of service to the upper classes in order to function as a redeemer of the popular will.

The military becomes concerned with internal security when a national revolution becomes a class revolution. Since most Third World societies are highly stratified, according to class, race, caste or area, the possibilities, even inevitabilities, of class conflict persist. Thus the military establishment, whether it so desires or not, is compelled to align itself with either traditional social classes or modern social classes. In a concrete sense, the role of the military in maintaining internal security is impaired by the very existence of class forces, for it is subject to pressure from both sides. It is possible to overthrow military establishments as part of a general upheaval against traditional class. At the same time, if the military is identified too closely with the popular classes, it tends to bring about an oligarchically inspired counterrevolution.

The role of the military is impaled upon the horns of a structural dilemma. In the very act of serving as an instrument of national redemption, it finds itself aligned against traditional class forces that have a great deal to lose in terms of wealth as well as prestige. At the same time, the use of the military as an instrument of suppression for riot control and secret police action has the effect of aligning the military against the popular class forces it is ostensibly serving. This problem is more acute in the Middle East and in Latin America than in most parts of Asia or Africa, where the military has not been replaced in the course of revolutionary action so much as it has found its roles transformed in the course of the developmental process (Rustow, 1963, p. 11; Vatikiotis, 1961). Interestingly, these former areas also contain nations where there is high tension and low stability.

There is an obvious tension between the function of military systems in the Third World that move toward socialism and populism and those that move toward more traditionalist economies and polities. The military clearly stands closer to the class and sector that actually wield power; and only when such classes and sectors are in a condition of impotence or cleavage does the military perform overt leadership tasks. One might say that within the "iron law of militarism" the choice is to be made between the military defense of national interests or the military definition of what these interests are. This is a more realistic goal than nonmilitary solutions in nations and cultures in which politics has been and continues to be militarized. The case of the Arab nations in this connection is most significant. The analysis by Avneri (1970, pp. 31–44) of militarism in North Africa deserves serious attention: "Within the Arab context, the emergence of military regimes has not signified a breakthrough to modernization, but a reversion to the traditional, legitimate form of government accepted and revered by the Arabs and by Islam for the past fifteen hundred years." Avneri (1970, p. 33) concludes the point by noting that "within the political culture of Arab society, military power equals political legitimacy, and for this reason military leaders have had very little difficulty establishing their authority in Arab countries" (see also Perlmutter, 1967).

The political leadership in the Third World tends to be drawn from the military corps. Therefore, in the "Egyptian socialism" of Nasser, or in the "guided democracy" of Sukarno, there was extreme emphasis on providing adequate policy-making roles for the armed forces. The tendency of political leadership in the Third World to simulate military models, and at times to adopt even the dress, manners and bureaucratic norms of the military establishment, underscores the close kinship and partnership between military and civilian elements.

The military often functions as a counterweight to the party apparatus on the one hand and the bureaucratic apparatus on the other. These two major forces of domination in the Third World are often kept harnessed and even bridled by the military acting in this adjudicatory role. The military acts to insure a partnership between party and bureaucracy, so that if the conversion from colonial rule to independence is not particularly constitutional, it is at least orderly. Involvement by the military in the political system and in the bureaucratic system can take place by promoting developmental programs or by consolidating or stabilizing popular revolution. But, whatever the mechanism, the fact of its involvement in the maintenance of the social structure is clear.

The strength of the military is often in inverse proportion to the absence of strength on the part of the middle or working classes. This is particularly true in the new nations. When the popular classes are too ineffectual in changing obviously bankrupt social relations, the elite of the armed forces perceive themselves as capable of filling a social vacuum. The army, by virtue of its national liberation character, may not have the capacity to crush opposition, but it may prevent any attempts to restore the old regime. It is remarkable how few counterrevolutionary or restorationist regimes have been successful in the Third World despite the large number of *coups d'état*. For although there has been an enormous amount of turmoil and transformation within the single party states during the post-colonial era, these have not led either to the restoration of colonial rule anywhere in the Third World, or to a particularly more advanced stage of social revolution once the nationalist phase has been achieved. However, in such multiparty states as the Congo, where there are no fewer than two hundred political parties and three different military directorates, there is far more instability than in the one-party political-military condition. Many leaders who were in charge when independence was attained in Africa remained heads of state or of the party a decade and even two decades later. Among these are Nyerere, Toure, Kenyatta, Senghor and Banda. Even when this original leadership collapsed, as in Algeria and Ghana, the new leaders were chosen from an alternative faction of the original revolutionary group—usually from the military faction of that early leadership. This indicates how ably strains in the social system are managed by the present power hierarchy.

The military in the new nations can oftentimes exercise international power. The Egyptian and the Algerian military see themselves very much concerned with problems of the unification of all North Africa, including the sporadic support of Palestinian Arabs urging the reconquest of Israel. The same kind of regional pattern is to be found with Indonesia in Southeast Asia, where it too hopes to function as a homogenizing force in the whole of the area, especially in the Malaysian peninsula. But thus far, the international role is more regional than truly worldwide. The internal role predominates.

Nevertheless, the search for regional, if not international roles cannot simply be dismissed as artificial. The annual regional meetings within the Third World provide a show of force no less than a show of political principles. The collective military might of the Third World determines its strength in relation to the First and Second Worlds. At the same time, the individual might of each nation within the Third World determines its position in relation to the other nations within that world. Even nations that are debtors before the International Monetary Fund may act as creditor nations within regional blocs. Thus, a nation like Egypt may be a borrower of funds, but it is a distributor of arms to other Middle Eastern nations. As long as the First World and Second World were organized against each other for the purpose of making nuclear exchanges, the Third World could not be considered militarily significant. With the widespread acceptance of war as a game, the rise of insurgency and counterinsurgency, and the nuclear standoff created by First and Second World competition, the conventional military hardware of the Third World has come to function as a more significant variable in international geopolitics than it did a decade ago.

When there is a balance of nuclear terror between the First and Second Worlds, relatively powerless Third World countries can become influential. However, one must be cautious on this point, since in fact such a perfect balance rarely exists. What does exist is a modernized version of the sphere of influence doctrine that still leaves the Third World in a relatively powerless position, at least with respect to exercising a role in foreign affairs. For example, when the Soviet Union decided to crack down on the revisionist regime of Czechoslovakia in the late 1960s, or equally when the United States decided to break the back of radicalism in the Dominican Republic a few years earlier, they were able to do so with impunity in part as a consequence of the unwritten sphere of influence doctrine established between the United States and the Soviet Union during the 1960s. Conversely, the prestige of the Third World, at least as a homogenous military unit, has been seriously undercut, for example, by the failure of the Organization of African Unity (OAU) to be able to resolve civil war in Nigeria, white rule in Rhodesia, or the continued colonial domination of Mozambique. The same situation obtains with other regional Third World clusters, such as the Organization of American States (OAS), which was virtually powerless during the Dominican crisis to affect United States foreign policy. In other words, when there is, in fact, a delicate equilibrium between the First and Second Worlds, then the military of the Third World can be important. Otherwise, it tends to be far less significant than, in fact, it would like to become (Haas, 1969, pp. 151–158).

The character of the military of the Third World is often shaped, symbolically, by the military powers that trained the armed forces or occupied the territory during the colonial period. Thus, in Indonesia, one finds a Japanese and Dutch combination; in Egypt, there are English and German types of models; in many parts of Latin America, there is a combination of French and German as well as United States prototypes. In other words, the actual organizational charts which describe these military organizations are based upon prototypes brought over from the former imperial power.

Even though the military in the Third World may think of itself as distinct and distinctively nationalistic, it still carries on the traditions of the old colonial armies. This is not just a cultural inheritance but a consequence of the complex nature of modern warfare, especially the complicated technology of advanced combat and the problem of training human forces so that they

become a significant military asset. Thus, there is rising tension between the
need for autonomy and the necessity for seeking out models and materials
from the advanced blocs. One way Third World nations attempt to overcome this contradiction is by the process of "spin-off," by relying on one
major power for technological and military hardware and on another for its
military organization charts. Many African nations, for example, Ghana, Nigeria, and Sierra Leone, which exhibited total military dependence on the
British style at the outset of the independence period, have moved to coun-
teract this by arrangements with advanced countries as disparate as West Germany, France and Czechoslovakia. At this point, a three-person rather than
a two-person game model comes into its own. The First and Second Worlds
increasingly must respond to Third World pressures, and not just the other
way around. For example, Great Britain must provide arms to Nigeria in
order to forestall the Soviet Union in the area, and France must refuse to
deliver aircraft to Israel in order to maintain Arab support and also Arab oil
supplies. However, as in the three-person game model itself, the outcomes can
often be frighteningly unpredictable, and, as in the case of the Nigerian civil
war or in the Indian-Pakistan war, gruesome.

Another reason for the diversification of military programming derives
from local factors. Thus, the strength of Ethiopia compels Somalia to accept
Russian military missions. Pressure from China leads India to diversify its
forces by using both Soviet and U.S. tactical and strategic weapons. A limiting factor is that the debtor nation faces the same problems in military terms
that it does in economic terms. The credit line is not irrevocable. This serves
to place an economic impediment and limit to the problem of political management in the Third World. Military assistance becomes a focal point in
maintaining exclusive relations with the former imperial power. For example,
the fact that France remains the exclusive distributor of military hardware
to most of its former African colonies can undermine the purpose of independence as much as it is undermined by exclusive trade arrangements (Gutteridge, 1955, pp. 117–129).

There is less apparent ambiguity in the military's role on the international
level than in its role as an internal agency. Military organizations directly
influence social structure by their allocation and distribution of power. The
military decides how much violence should be used in any internal situation.
Not only is the military reluctant to compete with civilian authority, but also
positive factors contribute to this situation. First, institutions subject to comparison, such as an army, are tested against other armies. The army can never
be wholly judged in terms of maintaining internal security. The military
thrust cannot completely avoid international considerations. Second, because
of the long-range aspect of foreign affairs, armed forces are immune to pragmatic tests of economic efficiency. They are not subject to the pressures of
private enterprise or to the rules of business and investment. They can function as a planning agency even within a "free enterprise" social order. Third,

an armed force generally has a style of its own. It is not subject to or limited by ordinary standards of behavior or legal canons (Pye, 1959, pp. 12–13).

These distinctive features also have negative by-products. The role of the army as an international agency requires a highly professional group of men, capable of making and rendering decisions on strategic issues, whereas an armed force concerned with internal security requires a much more political-ized orientation in which considerations of bureaucratic efficiency or separa-tion may be secondary. Hence, the multiplicity of roles for military estab-lishments in the Third World may involve structural incompatibilities. As Andrzejewski (1954, pp. 83–84) indicates, "Modern military technique pro-duced two contrary effects. On the one hand, it strengthened the centripetal forces by making subjugation of distinct regions easier; but on the other hand, it fostered a disintegration of multination empires, because universal conscript became an unavoidable condition of military strength, and armies raised in this way were of little value unless permeated by patriotism."

Democracy and Development

A continuing source of concern for those working in the area of compara-tive international development is the ever-widening disparity between politi-cal democracy and economic development. This disparity has been dealt with usually in what might be termed a "necessitarian" framework by scholars like Heilbroner (1963) who assert that a choice has to be made between political democracy and economic development. In point of fact, what is really being claimed is that no choice exists at all. Economic growth is a necessity, whereas political mobilization is declared to be a luxury. The mark of sovereignty, the mark of growth is, in fact, developmental, and therefore, is said to offer no real volition and no viable option.

On the other hand, there are economists and sociologists like Hoselitz (1960) and Moore (1967) who have taken what might be called a "liber-tarian" point of view. They claim that the disparity between democracy and development is real enough but that the costs of development remain uni-formly too high in the developing regions; and therefore, it is more important to preserve and enhance democracy than any irrational assumptions that de-velopment is either a singular mark of sovereignty or the necessary road to economic development. Underlying such thinking is a political determinism no less dogmatic, albeit far neater to advocate than the economic determin-ism under attack.

By now this argument of the 1960s has a somewhat arid ring. The assump-tions are either that development is necessary and we can do nothing about the costs involved, or that democracy is a categorical imperative and we must curb whatever developmental propensities we have to preserve this supreme good. Rather than penetrate the debate at this level, since the premises are either well known or well worn by this time, we should perhaps turn the matter around and inquire what has, in fact, been the relationship between political democracy and social and economic development over the past decade; and what are the dynamics in this Third World network of interrelationships?

The first observation that has to be made is that the process of comparative development includes a wide and real disparity between democracy and development. There exists a relatively high congruence between coercion and even terrorism and development, and a far lower congruence between consensus and development. In part, our problem is that middle-class spokesmen of the Western World have often tended to identify a model of consensus with a model of democracy, and both become systematically linked to what has taken place in North America and western Europe in the last 150 to 200 years: that is, a model of congruence in which political democracy and economic growth move toward the future in unison. The very phrase "political economy of growth" gives substance to this "bourgeois" model of development. When faced with the necessity of playing a role in the Third World and performing certain activities economically, politically, and socially in terms of the inherited model, what is retained in the rhetoric no longer can be sustained in performance. What one is left with, and why it has been difficult to confront so much theorizing on development, is a democratic model at the rhetorical level that is different from the capitalist model at the functional level. Further, there is a strong propensity, once this ambiguous model is accepted, to avoid coming to grips with the role of high coercion in achieving high development.

If we examine the available data, and here I shall restrict myself to the nonsocialist sector, in part, because there is a problem of data reliability, and also, because socialist systems have their own peculiar dynamic in relation to development and democracy, which requires a different set of parameters to explain relationships. Further, since there is a bias already established which links socialist systems to a high degree of political coercion, it is instructive to see how capitalist systems stand up to such coercive strains. Developmentalists are already prone to employ a Stalinist model as the basic type of socialist option. Thus, there is no problem in conceptualizing the relationship between coercion and development as a natural one when it comes to the socialist sector. However, when we turn systems around and look at the Western or capitalist sector of the developmental orbit, which reveals an absence of the same kind of coercive model, there are great problems in conceptualizing Third World tendencies.

When the word *democracy* is herein referred to it will be defined in terms of: (1) multiparty operations (2) under civilian regimes. Those two variables are key. There is no point in cluttering matters up with rhetorical theorizing about nice people who do good deeds. Simply put, democracy refers to multiparty control of politics on one hand, and civilian bureaucratic administrative control on the other. This definition is bare-boned and obviously subject to refinement. However, when talking about a military leadership or a military regime, we can be equally simple (hopefully not simple-minded). Military government ranges from outright rule of the armed forces without any civilian participation to co-participation by civilians under military domination and control. It will usually signify a single party structure, rather than a multiparty structure in the legal-superstructural aspects of political life.

The data herein examined are drawn from reports issued by the Organization for Economic Cooperation and Development (Martin, 1969). The figures concern growth rates of the total and per capita GNP output between 1960 and 1967 on average per annum; and per capita GNP in 1968 for selected developing countries. The information, although provided randomly, does break down into three large clusters. (1) There are those countries that are single party, under military rule, that have high developmental outputs and a high GNP rate over the decade. (2) There are, at the opposite end of the spectrum, those countries that are democratic (or relatively democratic) and have low GNP levels. (3) There is a clustering in the middle of approximately twenty nations that do not reveal any consistent pattern in terms of problems of conflict and consensus in development. It is not that they violate any model construction, but that they remain undecided —significantly uncommitted at economic levels and in political techniques for generating socioeconomic change.

Militarism and Development

Let me outline three national clusterings and see what the results are at the factual level. Such material is a useful place to start in this most dismal science called development (Martin, 1969, pp. 5–12).

In the high developmental, high militarization cluster, there are the following nations: Israel, Libya, Spain, Greece, Panama, Nicaragua, Iraq, Iran, Taiwan, the Ivory Coast, Jordan, Bolivia, Thailand and South Korea. Even a surface inspection indicates that this is hardly a line-up of democratic states. Let us directly examine the GNP figures, so that some sense of the extent of the aforementioned correlation can be gauged. On an annual percent increase over the decade, the percentile figures are as follows: Israel, 7.6; Libya, 19.2 (there are some special circumstances related to oil deposits in Libya, but, nonetheless, the figure is impressive); Spain, 5.9 (an interesting example because it is a long-militarized European mainland country); Greece, 7.5 (with no slow-up in sight under its present military regime); Nicaragua, 7.5 (one of the most "backward" countries in Latin America from a "democratic" point of view); Iraq, 6.9; Iran, 7.9; Taiwan, 10.0 (a growth figure which even the Soviets have recently marveled at); Ivory Coast, 7.5 (by all odds, one of the most conservative regimes in Africa and boasting the highest growth rate on the continent in the sub-Sahara region); Jordan, 8.8; Bolivia, 4.9 (a long way from the old socialist days of high foreign subvention);

Thailand, 7.1; South Korea, 7.6. This is a most interesting line-up. One would have to say that high development correlates well with high authoritarianism. Whether this is because authoritarianism quickens production, limits consumption, or frustrates redistribution is not at issue. The potential for growth under militarism remains an ineluctable fact.

Let us turn to the other end of the spectrum. These are relatively low GNP units: Venezuela, Argentina, Uruguay, Honduras, Ghana, Guatemala, Brazil, Dominican Republic, Senegal, Ecuador, Tunisia, Paraguay, Morocco, Ceylon, Kenya, Nigeria, Sudan, Uganda, India, Tanzania. Within a Third World context and without gilding the lily, and admitting that there are exceptions in this list like Paraguay, this second cluster in the main represents a far less militarized group of nations than the first list presented. It is instructive to list their GNP per capita annual percent increase: Venezuela, 1.0; Argentina, 1.2; Uruguay, 1.0 (which is one of the most democratic, one of the most liberal countries in South America); Honduras, 1.8; Columbia, 1.2; Ghana, which exhibits no percentile change over time in the whole decade; Guatemala, 1.9, Brazil, 1.2 (increasing, however, under military rule from 1964 to 1968); Dominican Republic, 0.7; Senegal, 1.2; Ecuador, 1.1; Tunisia, 1.5; Paraguay, 1.0; Morocco, 0.3; The Philippines, 1.0; Ceylon, 1.3; Kenya, 0.3; Nigeria, 1.6; Sudan, 1.2; Uganda, 1.2; India, 1.5; Tanzania, 1.2. The data plainly yields that the low rate of development intersects with the nonmilitary character of political mobilization in this second group of nations. Low militarization and low development are only slightly less isomorphic than high militarization and high development.

There is a most important middle group of nations not subject to this correlation coefficient. They do not share the same series of extensive polarities in GNP that vary in opposite directions in a perfectly consistent linear fashion. Chile, for example, has 2.4; Jamaica, 2.1; Mexico, 2.8; Gabon, 3.2; Costa Rica, 2.4; Peru, 3.2; Turkey, 2.7; Malaysia, 2.5; Salvador, 2.7; Egypt, 2.1; Pakistan, 4.1; Ethiopia, 2.7. Many of those nations have relatively stable GNP figures over time and do not easily fit the description of being depressed or accelerated in GNP rates. They are also the most experimental politically —at least during the 1960s. Certainly, experimentation (both by design and accident) characterizes countries like Pakistan, Egypt, Costa Rica and Chile. It is not entirely clear what this middle cluster of nations represents, or whether these trends are politically significant. Yet, they do represent a separate tertiary group and should be seen apart from the other two clusters of nations.

The critical level of GNP seems to be where levels of growth are under 2 percent, and where there is high population growth rate which more or less offsets the GNP. Under such circumstances, it is extremely difficult to achieve basic social services for a population, or maintain social equilibrium. For example, in India, if there is a 1.5 level of growth of GNP and a 2.4 level of population growth per annum, there is an actual decline in real growth rate.

This is how economists usually deal with the measurement of development. It may be faulty reasoning to accept *ex cathedra* this economic variable as exclusive; yet this measure is so widely used that the GNP provides a good starting point in our evaluation.

The correlations at the lower end of the model are not as good as at the upper end. Statistically, it is necessary to point out that militarized societies like Argentina and Brazil are not doing all that well economically. However, Latin America has a kind of "benign" militarism; a genteel quality that comes with the normalization of political illegitimacy. Thus, Latin American military regimes, in contrast to their African and Asian counterparts, have many exceptional features that account for why the fit of the model is better at the upper end of the GNP than at the lower end.

Before interpreting such information it should be noted that many nations clustering in the middle also seem to be the carriers of such experimental political forms as single-party socialism, communal living, socialized medicine and the like. The "ambiguous" nations also reveal an impressive movement toward some kind of democratic socialism that somehow eludes electoral definitions. In other words, the experimental forms seemed to be clustered in that middle grouping, whereas the nonexperimental nations tend to be polarized, just as the GNP itself is polarized.

The most important single conclusion is that the political structure of coercion is a far more decisive factor in explaining GNP than the economic character of production in any Third World system per se. Without becoming involved in a model of military determinism, the amount of explained variance that the military factor yields *vis à vis* the economic factor is much higher than the classical literature allows for. If we had comparable data for the socialist countries, and if we were to do an analysis of the Soviet Union over time, then we would see that there is a functional correlation between the coercive mechanisms that a state can bring to bear on its citizenry and the ability to produce high economic development, however development be defined, and ignoring the special problems involved in definitions based upon the GNP. For example, one problem of the GNP is the development of a cost-accounting mechanism whereby education is evaluated only by cost factors in input rather than output, whereas in goods and commodities you tend to have a profit margin built into the GNP figure. But such variations are true across the board; therefore, special problems involved in using the GNP formula are cancelled out in the larger picture. But the main point is that the element of coercion is itself directly linked to the character of military domination, while the specific form of the economy is less important than that relationship between military coercion and economic development.

Those critics in the West who celebrate progress as if it were only a matter of GNP cannot then turn around and inquire about the "quality of life" elsewhere (Hunt, 1966, pp. 134–156). Developmentalists cannot demand of foreign societies what they are unwilling to expect from their own society.

Too many theorists of modernization ask questions of the quality of life of countries that themselves ask questions about the quality of life. The military is the one sector, in most parts of the Third World, that is not absorbed in consumerism and commodity fetishism. The military is not a *modernizing* sector, but rather a developmental sector. Insofar as the military is autonomous, its concern is nation building: highway construction, national communication networks, and so forth. It creates goals that are not based on the going norms of commodities that characterize the urban sectors of most parts of the developing regions. The functional value of this model is that the linkage between the military and the economy is unique. The military is the sector that dampens consumerism and modernization and promotes, instead, forms of developmentalism that may move toward heavy industry and even heavy agriculture. This is a critical decision in nearly every part of the Third World. Because the military most often will make its decision on behalf of industrialism rather than modernism, it generates considerable support among nationalists and revolutionists alike. This is a big factor in explaining the continuing strength of the military in the Third World.

The data clear up a number of points. They help explain why many regimes in the Third World seem to have such a murky formula for their own economy or polity. For example, despite the brilliance of Julius Nyerere (1968, pp. 9–32) it is exceedingly difficult to determine the political economy of Tanzania. One reason for this is that there is a powerful military apparatus in nearly every expanding nation of the Third World. And this military structure, if it does not share in national rule directly, is directly plugged into the nation as an adjudicating voice between the political revolutionary element and the bureaucratic cadre.

To appreciate the role of the military in developing nations, we must go beyond the kind of economic definitions that have been employed either in Western Europe or in the United States. The murkiness of the Third World at the level of economics is, in fact, a function of the lack of clearly defined class boundaries and class formations. A critical factor is not so much social structure as social process. There is stability over time in these regimes. The political regime, the civilianized political regime, tends to be much less stable than the military regime. This is slightly obscured by the fact that in many of these nations there are *coups* within the *coup*—that is, *inner coups* within the military structure that function to de-legitimize civilian rule altogether. But these processes do not obscure the main fact that the military character and the military definition are not altered. For this reason, the relationship between military determinism and high economic growth tends to be stable over time; precisely to the degree that civilian mechanisms are found wanting.

Cutright and Wiley (1969–1970) came to this conclusion through an entirely circuitous and different route. They examined a mass of information on health, welfare, and security and found that the contents of the national political system usually become stabilized at that point in time when basic socioeconomic needs are satisfied. Further, there is no mass mobilization beyond such a point in time. If socioeconomic needs are satisfied within a socialist regime, then the Soviet system is stabilized. If such social needs are satisfied during a capitalist regime, then the capitalist system becomes permanent. If they are satisfied during an outright military dictatorship, then outright military dictatorship becomes normative and durable. In other words, the satisfaction of basic social services and economic wants is a critical factor beyond which masses do not carry on active political struggle. This quantitative support for the Hobbesian thesis on social order in the Leviathan has not been lost on the leadership of Third World nations, who continue to see the military as a stabilizing factor in economic development.

Political Stability and Development

If this foregoing analysis is correct, and if the military is able to solve these outstanding problems at the level of the GNP, we should be able to predict its continued stability. Since basic social services will be resolved at the particular level of military rule, political struggles will cease to assume a revolutionary character in much of the Third World. In point of fact, the character of socialist politics does not determine the strength of political behavior, but rather the other way around. The critical point in the Soviet regime comes at a time of Stalinist consolidation. During the 1930s, the Soviet regime achieved internal stability. The contours of socialism in Russia were thus fixed, some might say atrophied, at this specific historical juncture. True, certain adjustments have had to be made, certain safety valves have had to be opened to prevent friction or crisis. The Soviet Union has exhibited a move from totalitarianism to authoritarianism. However, basically the political organization is set and defined. There are no opposition movements in the Soviet Union, since there are no mass movements at the level of social discontent.

Similarly, in the United States, the point of resolution was when the political democracy became operational. Therefore, it continues to be relatively operational two hundred years later, even though great pressures have been brought to bear on the federalist system in recent times. Many Third World areas are stabilized at that point when military intervention occurs. When the initial revolutionary leadership vanishes (or is displaced) and when the bureaucracy and the polity are both bridled and yet oriented toward common tasks, at that point the military becomes powerful. This is also the moment when economic growth charts start rocketing upward. Therefore, it is no accident that this is also a moment at which fervor for political experimentation declines. What we are confronted with is not simply transitional social forms, but permanent social forms.

The trouble with most general theories of development is that they postulate conditions in the Third World as transitional when they fail to coincide with preconceived models. We are told, in effect, that military regimes in the

Third World are a necessary transition of "political economy" (Ilchman and Uphoff, 1969, pp. 3–48). Socialist doctrines of development often employ the same teleological model of politics for explaining away uncomfortable situations. Marxists declare that everything is in transition until achieving the height of socialism. Everything else is either an aberration, deviation or a transition. The difficulty with teleological explanations is that they work from the future back to the present, instead of taking seriously the present Third World social structures and political systems.

From an empirical perspective, social science determines what is meant by stability over time in terms of survival rates. Therefore, on the basis of this kind of measurement, the kind of network that exists in the Third World is, in point of fact, stable. What we are dealing with in many Third World clusterings are not permutations, or the grafting on of parts of other social systems. Third World systems are not transitional or derivational. They have worked out a modality of their own. Let us proceed one step further in the characterization of militarism and modernism. Going over the basic data presented, we can observe that what has happened is that the Third World, on the whole, and in particular those nations that exhibit a pattern of high economic growth during military rule, have accepted a Leninist theory of the state, while at the same time having rejected a Marxist theory of economics. That is to say, they accept the need for political coercion as a central feature of Third World existence, but at the same time deny socialist principles of economic organization.

The military state model is invariably a one-party model. Interparty struggle and interparty discipline vanishes. The party apparatus becomes cloudy as to its gubernatorial, bureaucratic and even political functions. The Leninist model is decisively emulated. The party serves the nation, but with a new dimension: The party also serves the military. The traditional Leninist model has the military politicized to the point of serving the ruling party. In the Third World variation, the political elites become militarized to the point of serving the ruling junta. And this role reversal of military and political groups is a decisive characteristic of the Third World today.

The economic consequences of this sort of Third World neo-Leninism is an acceptance of some kind of market economy based on a neocapitalist model. This process might be called: one step forward, and two steps backward. Any series of national advertisements will point out the low-wages, obedient-worker syndrome of many new nations. The model being sold to overseas investors by Third World rulers is, in large measure (whatever the rhetoric socialism may dictate), a model built on production for the market, private consumption, private profit, and a network that in some sense encourages the development of differential class patterns within Third World nations. But to gain such capitalist ends, what seemingly is rejected is the political participatory and congressional model common in northern and western Europe and the United States. There is no reason why this sort of system is transitional. On the contrary, given the conditions and background of underdevelopment, the kind of revolution made, the historical time of these revolutions and the rivalry between contending power blocs, this neo-Leninist polity, linked as it is to a neocapitalist economy, is a highly efficacious, functional and exacting model of the way most societies in the Third World have evolved in terms of political economy.

It might well be that with a greater amount of accurate data, this theory of the militarization of modernization will require modification, or even abandonment. However, the *prima facie* evidence would seem to indicate otherwise. More countries in the Third World have taken a more sharply military turn than anyone had a right to predict on the basis of prerevolutionary ideology or postrevolutionary democratic fervor. Therefore, the overwhelming trend toward militarism (either of a left wing or right wing variety) must itself be considered a primary starting point in the study of the Third World as it is, rather than how analysts might want it to become.

This is not a matter of either celebrating or criticizing the good society, or how Third World nations have fallen short of their own ideals. Few of us are entirely happy with any available system of society. Being critical of systems of society is a professional and occupational hazard. It is, however, not something that one would simply use as a proof that the system is unworkable. Indeed, if anything is revealed by the foregoing analysis, it is that the Third World has evolved a highly stable social system, a model of development without tears that forcefully draws our attention to the possibility that the Third World, far from disappearing, far from being transitional, far from being buffeted about, is becoming stronger, more resilient, and more adaptive over time. The widespread formula of military adjudication of political and bureaucratic strains within the emerging society is an efficacious model for getting the kind of mobilization out of "backward" populations that at least makes possible real economic development. It might well be that the strains in this military stage of development become too great, and the resolution itself too costly, to sustain real socioeconomic stability. However, at that point in the future, the Hobbesian laws of struggle against an unworkable state will once more appear and we shall know the realities of the situation by the renewed cries of revolution: This time against internal militarists rather than external colonialists.

REFERENCES

Andrezejewski, S. (1954) *Military Organization and Society.* London: Routledge & Kegan Paul.

Avneri, S. (1970) "The Palestinians and Israel," *Commentary,* vol. 49 (June), pp. 31–44.

Bondurant, J. V. (1958) *Conquest of Violence.* Princeton, N.J.: Princeton University Press.

Chambers, W. N. (1963) *Political Parties in a New Nation*. New York: Oxford University Press.

Cutright, P. and J. A. Wiley (1969–70) "Modernization and Political Representation: 1927–1966." *Studies in Comparative International Development,* 5 (Winter), 23–44.

Gutteridge, W. (1965) *Military Institutions and Power in the New States*. New York: Praeger.

Haas, E. B. (1969) *Tangle of Hopes: American Commitments and World Order*. Englewood Cliffs, N.J.: Prentice-Hall.

Halpern, M. (1963) *The Politics of Social Change in the Middle East and North Africa*. Princeton, N.J.: Princeton University Press.

Heilbroner, R. (1963) *The Great Ascent*. New York: Harper & Row.

Horowitz, I. I.. (1957) *The Idea of War and Peace in Contemporary Philosophy*. New York: Paine-Whitman. Republished (1973) New York and London: Souvenir Press, Humanities Press.

Hoselitz, B. (1960) *Sociological Aspects of Economic Growth*. New York: Free Press.

Hunt, C. L. (1966) *Social Aspects of Economic Development*. New York: McGraw-Hill.

Ilchman, W. F., and N. T. Uphoff (1969) *The Political Economy of Change*. Berkeley: University of California Press.

Janowitz, M. (1964) *The Military in the Political Development of New Nations*. Chicago: University of Chicago Press.

Kirk, M. (1964) *The Military in the Political Development of New Nations*. Chicago: University of Chicago Press.

Kirk, E. (1963) "The Rise of the Military in Society and Government: Egypt," in *The Military and the Middle East*. Columbus: Ohio State University Press.

La Palombara, J. (1963) *Bureaucracy and Political Development*. Princeton, N.J.: Princeton University Press.

Lipset, S. M. (1963) *The First New Nation: The United States in Historical and Comparative Perspective*. New York: Basic Books.

Martin, E. M. (1969) "Development Aid: Successes and Failures," *The OECD Observer,* no. 43 (December), pp. 5–12.

Meister, A. (1968) *East Africa: The Past in Chains, the Future in Pawn*. New York: Walker.

Ministry of Information and Broadcasting (1963) *India, 1963*. New Delhi: Government of India, Publications Division.

Moore, W. E. (1967) *Order and Change: Essays in Comparative Sociology*. New York: John Wiley.

Morris-Jones, W. H. (1960) "Mahatma Gandhi: Political Philosopher?" *Political Studies,* vol. 3 (February).

Naess, A. (1965) *Gandhi and the Nuclear Age*. Totowa, N.J.: The Bedminster Press.

Nehru, J. (1957) *Speeches: 1949–1953.* New Delhi: Government of India, Ministry of Information and Broadcasting, Publications Division.

Nyerere, J. K. (1968) *Freedom and Socialism* (Uhuru na Ujamaa). London and New York: Oxford University Press.

Perlmutter, A. (1967) "Egypt and the Myth of the New Middle Class," *Comparative Studies in Society and History,* vol. 19 (October), pp. 46–65.

Pye, L. (1959) *Armies in the Process of Political Modernization.* Cambridge, Mass.: M.I.T. Press.

Rustow, D. A. (1963) "The Military in Middle Eastern Society and Politics," in S. N. Fisher, ed., *The Military in the Middle East.* Columbus, Ohio: Ohio State University Press.

Vatikiotis, P. J. (1961) *The Egyptian Army in Politics.* Bloomington, Ind.: University of Indiana Press.

Werner, M. (1939) *The Military Strength of the Powers.* London: Gollancz; quoted from Klementi Voroshilov, *Fifteen Years of the Red Army.* Moscow, 1933.

PART II

United States

THE ROLE
OF THE UNITED STATES MILITARY
IN SHAPING FOREIGN POLICY

Adam Yarmolinsky

The world has changed a great deal in the last decade. Attitudes have changed, too—including my own. Several years ago, for instance, I thought I could find some justification for continuing the Vietnam War, once we had made the terrible mistake of letting ourselves become deeply involved.

Some things nevertheless remain the same. One thing that has not significantly changed is the role of the United States military, first, in shaping our Vietnam policy, and, now, in shaping our larger Southeast Asian policy. And that role, I submit, has been a negative one, intentionally or unintentionally negative regarding United States participation in Southeast Asian conflicts. Let me explain why I hold this perhaps somewhat astonishing belief.

In 1954, the Joint Chiefs of Staff told President Eisenhower that the only way we could intervene effectively in the battle of Dien Bien Phu, in which the French were about to be defeated, was with nuclear weapons, a wholly unacceptable alternative to the president. In 1968, the Joint Chiefs asked President Johnson for 206,000 more men, an equally unacceptable request, which produced an agonizing reappraisal of the whole Vietnam situation and the beginning of the turnaround, in the president's March 31 speech announcing the cutback in the bombing and, incidentally, taking himself out of the next presidential race. From 1964 to 1968, throughout the length of the war, the military discouraged its expansion and made its continuation more difficult. The military proposal to use nuclear weapons at Dien Bien Phu may have been the only practical alternative, or it may have been a sample of military rigidity and overkill. But its effect was to keep the United States out

of the war. The request for 206,000 more men in February 1968 may have
been intended to take advantage of what seemed a golden opportunity to make
up for cuts in U.S. military manpower outside Vietnam. But it helped to turn
the president and the country around.

Even in the earliest stages of our Vietnam involvement, the Green Berets
and the other U.S. military advisers were not popular with their senior military
colleagues, and the name of the "Never-Again" club was used to describe the
attitudes of American generals, many of them veterans of the Korean War,
towards another possible land war in Asia. In the early phases of the war, the
business-as-usual attitude of the military was a serious problem. The exigencies of the war interfered with military habits and priorities: the length
of the work day; and the concentration of research and development not on
what was needed in Vietnam, but on what was good for the long-run interests
of the military services. And one of the generally accepted reasons for the
cover-up of incidents like My Lai was the frequent rotation of troop commanders, so that every officer could "have his turn," whatever the effect on
military efficiency and effectiveness in the war.

The one thing that the military would not and could not do was to face up
to the near-impossibility of the tasks assigned them by their civilian masters.
In the process of not facing up, the military suffered terrible blows to its
command structure and to its morale. But I would argue that the false optimism of military reports about the famous light at the end of the tunnel
fooled no one in positions of real authority, except those who desperately
wanted to be fooled—as I think the Pentagon Papers have demonstrated.

So, it was not the military who pushed us into Vietnam or who kept us
there long after we should have departed. And Vietnam was not a special
exception. In the Dominican Republic, in Lebanon, in Korea, it was not pressure from the United States military that led to American intervention, however differently one may judge the merits of each case. It was, rather, a decision by the politically responsible civilian authorities in the United States
government, and a decision made essentially without consulting military authorities, except as to the military feasibility of the proposed course of action.

The use of military force in most cases was not pressed on the president. He
chose it. I refer in each case to a different president, and I would argue that
each president had his eyes open to the consequences in each case.

Why then, is there a general impression of the pervasiveness of military
influence on United States foreign policy? And what is the real nature and
extent of this influence?

Military influence seems pervasive, first because of the extraordinary size
and pervasiveness of the United States military establishment—an establishment constituting the largest institution in the United States and, indeed, in
the entire world.

It is difficult to convey some real sense of the size and the pervasiveness of
this enterprise. One way of describing it is that it represents roughly seven and

a half percent of the Gross National Product, or seventy-five cents out of every ten-dollar bill. Or one can think of it as something like eighty-five percent of all the money the federal government spends buying goods and services. Or as providing jobs for one out of every ten wage earners in the United States, not to mention the families that are dependent on those wage earners. This figure includes not only the people in military uniform, but also the civil servants and other employees of defense contractors who are engaged in the business of building weapons and weapons systems for the military establishment; and that's without counting the butcher, the baker, and the candlestick-maker who are dependent on the business of soldiers, sailors, airmen and marines, civil servants, and defense contract employees.

Where other departments of the federal government measure their size in millions of dollars, the Defense Department measures its size in billions of dollars. Where other departments speak of thousands of men, the Defense Department speaks of millions of men.

Not only is the defense establishment uniquely large; it is also uniquely extensive. It is present in every community in the United States where there is a military base or a defense plant, a defense contractor, a sub-contractor, or a sub-sub-contractor, or even where there is a recruiting station. Almost any place that merits a spot on the map has some connection with the Department of Defense, and the people who live there are very much aware of it.

That is one reason why, whether the facts support it or not, we have a general impression of the pervasiveness of military influence on U.S. foreign policy—and on everything else in the United States.

There is a second reason, or set of reasons that goes to the special consideration given, at least until recently, to the military budget, as distinguished from all the other budgets that go into the overall federal budget.

President Kennedy used to say that the United States could afford all of the defense spending that it needed to protect its security. On the face of it, that is a sort of self-evident commonsensical statement. But why did not President Kennedy, or other presidents, say that the United States can afford all the spending on education that it needs, to give the people of this country a decent education? Why did he not say we can afford all the money for welfare that is needed to overcome the problem of hunger and homelessness, inadequate health care, inadequate clothing, and inadequate access to the necessities of life? Why did he not say we can afford all the money for our federal housing programs that is necessary to do what Federal Housing Acts have been saying is the policy of the United States since 1937—a decent house in a decent neighborhood for every American family.

Somehow the military emerged as a special case, and they continue to be a special case. One of the reasons the military gets the kind of special treatment from the Congress of the United States is that there is a coincidence of interest between the industrial suppliers to the military market and the government servants, military and civilian, who operate the military establishment. That coincidence of interest is particularly powerful, because so many of those industrial suppliers are specialized to the needs of the military establishment. If one is in the business of making supersonic missiles, it is awfully hard to convert factory staff and executives to the business of making aluminum tooth picks. There aren't very many things that most defense contractors can do outside the defense market. They are looking very hard for what they can do right now, but there is a kind of specialization in the defense supply business which doesn't apply nearly so much in other kinds of businesses that supply other sectors of the federal, state, or local governments.

Because industrial suppliers and their employees are significant constituents of members of Congress, this coincidence of interest has supported and reinof the United States approach to the funding of national defense for the last fifteen or twenty years. There are some signs that it is beginning to change. I must emphasize that here again we are examining why we have the impression of the pervasiveness of military influence on foreign policy, without regard to the facts as they actually are. It is a long step from weapons acquisition policy, i.e., the way in which the military's budget is shaped, to the way in which our foreign policy is shaped. The special position and the special consideration given to the military budget do not necessarily mean the military will get special consideration when it expresses its views on questions of American foreign policy. So that we are still considering what makes it look that way rather than how much it is really that way.

The third factor that makes it look that way is the .vociferous quality of political expression by political extremists, in the military services or in retirement from the services, on questions of American foreign policy. By and large the American military have been followers rather than leaders in political expression. The ethic of non-participation in politics is still very strong. There is an old tradition among career military men of not voting, even in national elections. That tradition has begun to break down; it has still by no means disappeared. But when the military gets an idea or a concept from political leaders, they hang on to it, some might say like grim death, and they have hung on to the ideas of the cold war even after civilian leaders have accepted modifications of these ideas in a more pluralistic and complex world. Some of the programs that were set up at the height of enthusiasm about the cold war, like the traveling seminars of the Industrial College of the Armed Forces, or the Joint Civilian Orientation Conference, which takes civilians around the country and shows them the armed forces at work, have continued to foster these attitudes. The views of some atypical, but extremely articulate retired officers are still periodically in the news, although apparently they are not paid very much attention by those in high office—even by the present occupants of the White House.

Of course, the informational activities of the Pentagon tend to paint the activities of the military in the best possible light, just as the informational activities of any institution attempt to paint its activities in the best possible light. The trouble with the Pentagon is that it is so big that, even when it spends a relatively small proportion of its resources on public information or propaganda (whatever you want to call it), the results come out very strong and very loud indeed. So, my feeling that the Pentagon effort needs to be regulated and controlled doesn't apply to the way it's being done, but rather to the size of the effort itself.

There is one more reason in my list of reasons, why, whatever the facts are, one gets the impression that United States foreign policy is shaped by the military. The United States finds itself too often including repressive regimes among its friends and allies around the world. The list is well-known. But this is, I believe, one of the unhappy, and, I believe, unavoidable, consequences of our role as a great power. Again, in my view, it is at least as much a function of our State Department bureaucracy as it is of our military bureaucracy. In 1962, when Robert Kennedy took a round-the-world trip to visit the less developed countries of Asia, Africa, and Latin America, where he made it his business to talk with people on the streets, in their houses, and in the fields, he came back very much upset. He said, in effect, here we are in spirit the youngest country in the world, out on the New Frontier, anxious to experiment with new, fresh, and vigorous ideas, and we are regarded generally in the world as an old fogy. We have the wrong friends and the wrong enemies. We aren't trying to make friends with the rising generation of political leaders in these developing countries, the people who are now in the opposition—and may never be in the government at all. We are content to be friends with the status quo. That wasn't enough, he said, and, being Robert Kennedy, he organized an effort to do something about it. I had the good fortune to be involved fairly actively in that effort over the years from 1962 to 1964, and I think we made a little difference.

But in making that difference we were not battling the military bureaucracy. We were trying to shake up, to move, to inch along the State Department bureaucracy, all dressed up in their civilian clothes. So I submit that the fact that the United States too often finds itself allied with repressive, stand-pat, status quo regimes is no indication of military influence on our foreign policy.

Still, an institution as large, as powerful, and as pervasive as the military must have some significant influence on United States foreign policy. Clearly it has influence on other things in this country. Here I propose to suggest the ways in which we can identify and mark out the locus of military influence on U.S. foreign policy.

First there is the evident availability of military "solutions" to foreign policy problems. After all, if we have this enormous military establishment that is very big and very efficient and very good at getting memoranda and reports into the White House, ahead of the reports that are asked for from the State Department, the Department of the Treasury, or the Department of Agriculture, and if we are aware of the fact that here are all of these ready, flexible military forces, before we begin to look for other kinds of solutions to our problems, it is understandable that our political leaders ask themselves: How about using the military? Other kinds of solutions are likely to be more complex, more subtle, possibly even in the short run more expensive, because we have already paid the big bills for the military. Those large costs are fixed costs, while the variable costs of using them are relatively small.

Consider the terrible psychological effect that it must have had on those who had to decide what to do about the prison revolt in Attica, knowing that all those state troopers were in place and had their shotguns and rifles loaded, and that there was one sure way of "solving" the Attica problem. It might turn out, as they were later to discover, to be unpleasantly close to what the Germans call a final solution, but it was a ready solution.

In the case of the U.S. military, it wasn't just the presence of military forces, like the presence of state troopers at Attica, but it was also a certain mystique that attached in the early years of the sixties to the counterinsurgency forces like those of the Green Berets. It grew out of an understandable and natural human reaction to the fact that President Kennedy had abandoned that old frightening doctrine of John Foster Dulles, in the previous administration, called massive retaliation. President Kennedy had explicitly said we weren't going to use nuclear weapons at places and times of our choosing to support our foreign policy objectives. Once it had been established that the United States was no longer in the business of invoking Armageddon to support American foreign policy, everyone heaved a collective sigh of relief. Looking at the military forces at the other end of the spectrum of violence—counterinsurgency forces, counterguerrilla forces, Green Berets—they thought, almost with a kind of euphoria, that this was the sort of war that would be relatively safe to engage in, because it was not going to bring the end of the world in a great radioactive cloud.

Then, too, there was the fact that not only were the military forces on hand, but the military as a presence was on hand. The day after President Nixon was inaugurated, he was interviewed in his office by a reporter. His furniture had scarcely been moved in, had not yet been arranged, and the oval office was rather bare, but in one corner there was a display of flags: the Army flag, the Navy flag, the Marine Corps flag, and the Air Force flag. The Army flag had all the battle streamers attached to it, and they were all in the right order. Each one had the proper brass eagle on top of it, and it was mounted on a regulation flag stand. The reporter asked, "Mr. President, where did that come from?" and he replied, "Oh, They put it there." They did put it there, and They were somehow efficient enough so that They put it there before it occurred to anybody that perhaps the flag of the Department of Health, Education and Welfare should be installed in the office of the President, along with an encyclopedia, or somebody's book of home health cures. The military got there first, and they got there first in the way that made it seem perfectly natural. Of course, They put it there, and if They put the flags there, They may have put some ideas there at the same time, which again nobody quite realized had crept in.

There is another way of looking at this issue of availability of military solutions. What you cannot accomplish, you do not consider. If you don't have the military forces to pursue a particular course of action, you don't consider military action as an available option.

President Roosevelt delivered his famous "Quarantine the aggressors" speech four years before he felt even reasonably confident that the United States had the kind of military forces that could enter into World War II. At the time of the Cuban missile crisis, when we did invoke the United States Navy in a quarantine operation—about as limited a military activity as could be mounted under the circumstances—it became painfully evident that if such an operation had been attempted one year before, the Navy would have been ill-equipped to carry it out.

There is at least one special role in foreign policy-making that the military does have. It is not a role that exists at the seat of government in Washington, but rather in American diplomatic missions throughout the world, where military, naval, and air attachés, particularly in smaller missions in less developed countries, tend to make up what I think is fair to characterize as a disproportionate share of U.S. personnel in these missions. For example, in the Dominican Republic it was the military attachés who carried the news about the supposedly alarming developments that led to U.S. intervention in 1965. These attachés were naturally in touch with their opposite numbers, and inclined to be at least sympathetic professionally to them, and their information tended to overshadow the information that came from the civilians who were also collecting information for the embassy.

There is a qualifier on that aspect of military influence, because the critical importance of ambassadors diminishes as better communications and better transportation make the role of the man on the spot less important. Thus, the role of the military attachés in the field diminishes proportionately. Of course, military attachés are part of the whole military assistance machinery, a machinery by which we supply supposedly obsolete military equipment (which may, however, be declared obsolete in order to supply it) to friendly governments. I do not suggest that this is a kind of con game, but bureaucracies have a way of getting their things done, and the military assistance program is a system by which we are trying to buy friends in some parts of the world. Military assistance can be and has been a valuable instrument of United States foreign policy in maintaining a degree of stability in Europe, in Israel, in Korea; but, like any instrument, it can be over used. It takes effective controls outside the bureaucracy that established and operates the military system to prevent that bureaucracy from using it to trade favors among national military establishments. Again, the availability of military assistance, or even the availability of military credit sales (where the United States sells equipment to the recipient country on rather favorable credit terms which it could not get commercially) is an investment of foreign policy that may delude us into believing that we can control the situation outside the territorial limits of the United States, when it is really quite beyond our control.

Take the Pakistan/India conflict for example. The last time those two nations got into a fight, the United States was able to stop it, in effect, because we were the principal supplier of arms to both sides. When we cut off their supply of ammunition and spare parts, they ran out of war. But we are no longer the principal supplier, so we can't do that anymore. This time we tried to stay on the good side of the President of Pakistan by continuing to supply arms that ended up being used for genocidal killing in East Pakistan, and that killing may well have helped to trigger military reaction that set off the war itself. At that point not only did we continue to supply arms— but we also continued to pretend that we weren't doing it. When Senator Edward Kennedy was told no, there weren't really any continuing arms shipments, and it turned out that there were just arms shipments that hadn't quite been finished off after the order was given to stop them, so that they were going on but they weren't continuing, what he got was not a lie about those shipments; it was just the kind of bureaucratic answer that bureaucracy (in this case State Department bureaucracy) gives when it is not under effective control.

One more way in which the military can be said, in a sense, to influence U.S. foreign policy, goes to the atmospheric conditions in Washington as they affect the role of the military bureaucracy, vis-à-vis other bureaucracies. There is a kind of pecking order among the various bureaucracies that make up the federal government, and we have been putting the military consistently at or near the top of that pecking order. Parts of the pecking order are based on available resources. But here again the sheer brute size of the Defense Department enables it not only to spend dollars, but to provide limousines and drivers and private airplanes to move people who might have otherwise to travel by commercial airlines.

But in part the pecking order is based on the kind of value judgments that Americans make. How would you rate the United States Commissioner of Education against the Commanding General of the Strategic Air Command? How would you expect the two figures to be rated in the nation's capital? Or how would you rate the Social Security Commissioner as against the United States Commander in Vietnam? The issue is not one of personal judgment, but of the amount of relative respect that the political system accords. Bureaucrats notice these things; that is their business. Bureaucrats outside the military establishment believe they are less powerful than their military counterparts, and act accordingly. Perhaps this is so because they feel that society begrudges them their power more than it begrudges the power of the military bureaucrats. This kind of atmospheric condition in turn increases the influence of the military on foreign policy, in ways that are very hard to point out specifically.

In sum, military influence in United States foreign policy is a good deal less than we might suppose. If it is more than we would prefer, then I suggest, to make a bad paraphrase, that the fault lies not in the stars, nor the eagles, nor the oak leaves, nor the gold braid, but in ourselves.

THE EMERGENT MILITARY:
CIVIL, TRADITIONAL, OR PLURAL?*

Charles C. Moskos, Jr.

Academic definitions as well as ideological attitudes of the American armed forces fluctuate between two poles. At one end are those who see the military as a reflection of dominant societal values and an instrument entirely dependent upon the lead of civilian decision-makers. Conversely, others stress how much military values differ from the larger society and the independent influence the military has come to exert in civil society. In a real sense, these two emphases differ as to whether the armed forces or society is primary. Yet, neither conception is wholly wrong nor wholly accurate. Rather, the issue is one of the simultaneous interpenetration and institutional autonomy of the military and civilian spheres.

At the outset, nevertheless, it should be made clear that the conceptual question of the independent versus dependent relationship of the military and civilian orders is not intrinsically a value judgment. Indeed, we find diverse viewpoints on the conceptual question crisscrossing political positions. Thus, supporters of the military organization have argued both for and against greater congruence between military and civilian structures. Likewise, the harshest critics of the armed forces have variously claimed the military establishment to be either too isolated or too overlapping with civil society. The point here is that at some level a sociological understanding of the armed forces and American society can be analytically distinguished

* Reprinted with permission from *Pacific Sociological Review*, vol. 16, #2 (April 1973): 255–280. Support from the Russell Sage Foundation is gratefully acknowledged.

from a political position. In fact, of course, this is not always so readily apparent when one gets down to concrete cases. But, it is my personal statement that an ideological position must ultimately lead to social science analyses. This is especially mandatory in the present period when the American military establishment is undergoing profound changes both in its internal organization and in its relationship to the larger society.

Armed Forces and American Society in Recent Retrospect

Even in the single generation which has elapsed since the start of the Second World War, one can readily observe that the American military establishment has passed through several distinctive and successive phases. Prior to World War II the military forces of this country constituted less than one percent of the male labor force. Armed forces personnel were exclusively volunteers, most of whom were making a career out of military service. Enlisted men were almost entirely of working-class or rural origin, and officers were overproportionately drawn from southern Protestant middle-class families. Within the military organization itself, the vast majority of servicemen were assigned to combat or manual labor positions. Socially, the pre-World War II military was a self-contained institution with marked separation from civilian society. In its essential qualities, the "From-Here-To-Eternity" army was a garrison force predicated upon military tradition, ceremony, and hierarchy.

The Second World War was a period of mass mobilization. By 1945, close to twelve million persons were in uniform. Although technical specialization proceeded apace during the war, the large majority of ground forces were still assigned to combat and service units. Even in the Navy and Air Corps —services where specialization was most pronounced—only about one-third of personnel was in technical or administrative specialties. The membership of the World War II forces was composed largely of conscripted or draft-induced volunteers. To put it another way, the military of World War II, while socially representative of American society, was still an institution whose internal organization contrasted sharply with that of civilian structures. At home, nevertheless, there was popular support of the war and criticism of the military establishment was virtually non-existent.

Following World War II, there was a sixteen-month period when there was no conscription at all. By the time of the outbreak of the Korean War in 1950, however, the draft had already been reinstituted. The conflict in Korea was a war of partial mobilization, slightly over 3,600,000 men serving at the peak of hostilities. Organizationally and materially, the armed forces of the Korean War closely resembled those of World War II. Unlike World War II, however, the war in Korea ended in stalemate which in turn contributed to adverse accounts of soldiers' behavior: prisoner-of-war collaboration, the lack of troop motivation, and the deterioration of military discipline.

The cold-war military, which took shape after Korea, averaged around 2,500,000 men, again relying in great part on the pressures of the draft for manpower. Especially significant, technical specialization became a pervasive trend throughout the military during the 1950s and early 1960s.[1] The proportion of men assigned to combat or service units declined markedly with a corresponding increase in electronic and technical specialists. These trends were most obvious in the air force, somewhat less so in the Navy, and least of all in the Army and Marine Corps. Moreover, because of the post-Korea doctrine of nuclear deterrence and massive retaliation, it was also the Air Force which experienced the greatest proportional growth during the 1950s.

Although cold-war policies were generally unquestioned during the 1950s and early 1960s, the military did not escape embroilment in political controversy during the cold-war period. Such controversy was centered on issues of military leadership and the institutional role of the military. Command policies at the highest level were subjected to conservative charges in two major Senate hearings. The military establishment found itself on the defensive in countering charges of being soft on Communism in both the McCarthy-Army hearings of 1954, and in the 1962 hearings resulting from the cause celebre following Major General Edwin Walker's relief from command (for having sponsored troop information programs with extreme conservative content). During the same period, intellectuals on the Left emerged from their quiescent stance and began critically to attack the military establishment from another direction. Deep concerns about the military-industrial complex in American society were raised—an issue which was to achieve fruition of sorts over a decade later. By and large, however, the cold-war criticisms of the military were relatively weak and in basic respects the armed forces maintained the high regard of the American public.

The war in Vietnam ushered in another phase of the armed forces in American society. There was the obvious increase in troop strength, to a high of 3,500,000 in 1970. At the same time the role of the Air Force and Navy went into relative descendancy as the Army and Marine Corps came to bear the brunt of the conflict in Indochina. The Vietnam War also led to deviations from the cold-war policies of manpower recruitment. In 1966, entrance standards were lowered to allow the induction of persons coming from heretofore disqualified mental levels—overproportionately lower class and black. In 1968 the manpower pool was again enlarged, this time by terminating draft deferments for recent college graduates—largely middle-class whites. For the first time since the Korean conflict, the membership of the armed forces was again bearing some resemblance to the social composition of the larger society.

1. Kurt Lang, "Technology and Career Management in the Military Establishment," in Morris Janowitz, ed., *The New Military* (New York: Russell Sage Foundation, 1964), pp. 39–81.

If the debates concerning the military establishment were generally muted in peacetime, this was not to be the case once America intervened massively in Indochina. Although there was a brief spate of glory attached to the Green Berets, opposition to the war soon led to negative portrayals of the armed forces. As the antiwar movement gained momentum, it began to generalize into a frontal attack on the military system itself—particularly within elite cultural and intellectual circles. The 1967 March on the Pentagon crossed a symbolic threshold. Not only was the war in Vietnam opposed, but for a growing number the basic legitimacy of military service was brought into open question. Adding to the passion of the antimilitarists were the revelations of American atrocities in Vietnam, and the physical and ecological devastation being perpetuated throughout Indochina. To compound matters, there were a host of other factors somewhat independent of Vietnam which served to tarnish the image of the American military: the capture of the *Pueblo,* the inequities of the draft, reports of widespread drug abuse among troops, corruption in the operation of post exchanges and service clubs, astounding cost overruns in defense contracts, and military spying on civilian political activists.

Even more telling, there were undeniable signs in the late Vietnam period of disintegration within the military itself. Some numbers of men in uniform —white radicals, disgruntled enlisted men, antiwar officers—were increasingly communicating their feelings to other servicemen as well as to groups in the larger society. Moreover, throughout all locales where U.S. servicemen were stationed, racial strife was becoming endemic. The possibility that black troops might owe higher fealty to the black community than to the United States military began to haunt commanders. In Vietnam, the American military force by 1971 was plagued by breakdowns in discipline including violent reprisals against unpopular officers and noncoms. Although much of the malaise in the ranks was attributed to change in youth styles—as manifest in the widespread use of drugs, it was more likely that the military's disciplinary problems reflected in larger part that general weakening of morale which seems always to accompany an army coming to an end of a war. Even the use of sheer coercive power on the part of commanders has limitations once the esprit of an armed force has been so sapped.

The contrast in ideological and public evaluations of the American military establishment over three wars is revealing. In the Second World War, the American military was almost universally held in high esteem in a popularly supported war. Conservatives and isolationist sectors of American public opinion were quick to fall in line behind a liberal and interventionist national leadership. In the wake of the Korean War, defamatory images of the American serviceman were propagated by right-wing spokesmen. Liberal commentators, on the other hand, generally defended the qualities of the American armed forces. In the war in Southeast Asia a still different quality has emerged. Although initially an outcome of a liberal administration, the

war had come to be primarily defended by political conservatives while the
severest attacks on both the behavior of American soldiers and the military
establishment now emanate from the Left.

But even beyond Vietnam and factors unique to armed forces and society
in the United States, the decline in status of the American military establishment may well be part of a more pervasive pattern occurring throughout
Western parliamentary democracies. Observers of contemporary armed
forces in Western Europe, the United Kingdom, Canada, and Australia, have
all noted the sharp depreciation in the military's standing in these societies.
Indeed, although it seems somewhat far afield, the possibility suggests itself
that Vietnam may be a minor factor in explaining the lessened prestige of
the American military establishment. This is to say that the American military, like its counterparts in other Western post industrialized societies, is
experiencing an historical turning point with regard to its societal legitimacy
and public acceptance.

The Social Composition of an All-Volunteer Force

A stated goal of the Nixon administration was the establishment of an all-
volunteer force. Shortly after assuming the presidency, Nixon appointed a
commission to study the implementation of the all-volunteer force. This panel
—referred to by the name of its chairman, Thomas S. Gates—published its
report in February 1970. It was the unanimous recommendation of the Gates
Commission to establish an all-volunteer force with a standby draft. In July
1970, legislation was introduced in the U.S. Senate which would put into
effect the recommendations of the Gates Commission. Although yet to be
passed, the breadth of support for this legislation was revealed in its sponsorship which included political figures as diverse as Barry Goldwater and
George McGovern. Moreover, a bill was passed and signed by the president
in 1971 which substantially increased military salaries, especially for servicemen in their first tours. The Department of Defense set July 1, 1973, as the
goal for achieving a "zero draft." Under this plan, all entering servicemen
are volunteers with the proviso that Congress retains a two-year extension
of induction authority and standby authority.

It seems fairly certain, then, that, sooner rather than later, this country's
generation-old reliance on the draft for military manpower will come to an
end. Before looking at some probable consequences of this change, however,
some background data is in order on military procurement and retention
rates in the modern era. Over the past two decades, with some variation,
about one-third of all age-eligible men have failed to meet the mental test
standards required for military entrance; this group has been disproportionately poor and/or black. About a quarter of the age-eligible men have obtained draft deferments (primarily educational) which have resulted in de
facto exemptions; this group has been greatly overrepresentative of upper-

middle-class youth. (In the most recent period, upper-middle-class youth have also decreased their draft liability by utilizing liberalized conscientious objection procedures and obtaining medical documentation of ersatz physical disabilities.) Thus, only about forty percent of age-eligible young men actually have served in the military in recent years; and these men were over-proportionately drawn from the American stable working and lower-middle classes.

Between the wars in Korea and Vietnam, about one-fourth of all incoming military personnel were draftees (in almost all cases these were army entrants). About another quarter were draft-motivated volunteers, that is, men joining the military to exercise a choice in time of entry or branch of service. Therefore, only about half of all entering servicemen in peacetime were "true" volunteers, i.e., men who would have presumably joined the service without the impetus of the draft. During both the Korean and Vietnam Wars the number of draftees and draft-motivated volunteers increased sharply. It was estimated that in 1970 less than twenty-five percent of incoming servicemen were "true" volunteers.

Once within the military, retention rates vary by manner of service entry. In peacetime years, about twenty to twenty-five percent of volunteers reenlist for a second term; among draftees the proportion going on to a second term averages about ten percent. In the later years of the Vietnam War, however, volunteer reenlistments dropped to fifteen to twenty percent, and draftee reenlistments were less than five percent. Once a serviceman has made the transition from first to second term, however, he has usually decided upon a military career. With remarkable consistency about four out of five second-term servicemen remain in the military to complete at least twenty years service (the minimum time required for retirement benefits).

What lessons does the experience of the recent past offer for an understanding of the military establishment which will emerge from the institution of an all-volunteer force? Will the armed forces maintain a membership which resembles in basic respects that which existed prior to the Vietnam War, or will the social composition of the military undergo a fundamental transformation? Not too surprising, as in most controversial issues, social science data has been quoted to assert contrary predictions and conclusions.

Changes in the Enlisted Ranks. One of the most telling arguments against the establishment of an all-volunteer force is that such a force will have an enlisted membership overwhelmingly black and poor.[2] Yet the Gates Commission counters: "The frequently heard claim that a volunteer force will be all black or all this or all that simply has no basis in fact. Our research indicates that the composition of the armed forces will not fundamentally change by ending conscription. . . . Maintenance of current mental, phys-

2. Harry A. Marmion, *The Case Against a Volunteer Army* (Chicago: Quadrangle Books, 1971).

ical, and moral standards for enlistment will ensure that a better paid, volunteer force will not recruit an undue proportion of youths from disadvantaged backgrounds." [3]

Another study, contracted by the Institute of Defense Analyses, differs on virtually all counts from the findings of the Gates Commission.[4] Based on a detailed statistical comparison of civilian and military employment earning potential, the Defense Analyses report concludes: (a) non-high school graduates suffer a financial loss if they choose civilian employment over continued military service; (b) enlisted men who have attended college experience a financial loss if they remain in military service; (c) military and civilian earnings for high school graduates are roughly the same; and (d) military earnings for blacks with a high school education or less will far exceed their earnings in the civilian labor force. In other words, on the assumption that social groups will generally behave in their own economic self-interest, an all-volunteer force would significantly overdraw its membership from the less educated and minority groups of American society.

Reference to the experience of all-volunteer forces in other nations is also inconsistent. Again, in support of the all-volunteer force, the Gates Commission finds: "The recent experience of the British, Australian, and Canadian Armed Forces suggests that competitive wages will attract an adequate quantity and quality of volunteers." [5] Yet, an account of the British experience notes that typical recruits are "untrained school-leavers" coming from older and impoverished urban areas.[6] Indeed, over a third of British volunteers now join the armed forces *before* their seventeenth birthday (twenty percent of all British volunteers being only 15 years old!).[7]

With such contradictory findings, what are we to conclude as to the probable future social composition of the enlisted ranks? In all probability an all-volunteer force will be less socially representative than the present military establishment, but, with pay raises, nowhere near exclusively dependent on the lowest social and economic classes. That is, a reasonable expectation is that the rank and file of a non-conscripted military will fall somewhere between the claims of the Gates Commission and the dire predictions of an "all-black" or "all-poor" force.[8] Which of the two extremes

3. *The Report of the President's Commission on an All-Volunteer Armed Force* ("Gates Commission") (New York: Macmillan, 1970), pp. 15–16.
4. Gary R. Nelson and Catherine Armington, *Military and Civilian Earnings Alternatives for Enlisted Men in the Army*, Research Paper P-662 (Arlington, Va.: Institute for Defense Analyses, 1970). See also, K. H. Kim, Susan Farrell, and Ewan Clague, *The All-Volunteer Army: An Analysis of Demand and Supply* (New York: Praeger, 1971).
5. *President's Commission*, p. 168.
6. Gordon Lee, "Britain's Professionals," *Army*, July 1971, pp. 28–33.
7. Ibid., p. 31.
8. In this regard, a 1969 survey based on a representative national sample of high school male students found an amazingly high 16 to 25 percent who said they

an all-volunteer force will tend toward will be determined largely by the eventual total manpower strength of the armed services. A smaller force—say, close to two million persons—will be able to afford higher entrance standards, thus precluding overrecruitment from America's underclasses. Conversely, a larger force will have to draw deeper from previously unqualified groups.

Changes in the Officer Corps. The movement toward an all-volunteer force will be accompanied by significant changes in the social bases of officer recruitment.[9] The ROTC units from which the bulk of the officer corps is now drawn will almost certainly decrease in number and narrow in range. Partly as a result of anti-ROTC agitation at prestige colleges and universities, ROTC recruitment will be increasingly found in educational institutions located in regions where the status of the military profession is highest—rural areas and in the South and Mountain states. It must be candidly acknowledged that such ROTC units will often be at colleges and universities with modest academic standards. Within the larger urban areas themselves there is a possibility that ROTC units may be removed from campuses and instituted instead on a metropolitan basis. This eventuality would most likely further restrict recruitment of ROTC cadets coming from upper-middle-class backgrounds.

Moreover, the armed forces will obtain a growing proportion of its officers from the service academies, a step which has already been taken by substantially increasing the size of the student body at these institutions. Although the system of selection into the service academies is broadly based, there is a strong possibility that military family background will become even more prevalent among academy entrants. Because of the expansion of the armed forces over the past twenty years, the number of such military families and their offspring has increased markedly. Such excessive selection from military families—officers and enlisted—would result in a separation of the officer corps from civilian society by narrowing the basis of social recruitment. Likewise, any increased reliance on government-sponsored military preparatory or even privately-sponsored preparatory schools would similarly narrow the social and geographical background of future officers.

Finally, there is the probability that recruitment from the ranks into the officer corps will decline. With the greater and greater emphasis on a college degree, there will be an acceleration of the trend to recruit from college grad-

---

would volunteer for the armed forces, given no draft and no war. Jerome Johnston and Jerald G. Bachman, *Young Men Look at Military Service* (Ann Arbor, Mich.: Institute for Social Research, June 1970).

9. Most of the discussion given here on the probable changes in the social background of the officer corps in an all-volunteer force is a paraphrase of Morris Janowitz, "The Emergent Military," in C. C. Moskos, Jr., ed., *Public Opinion and the Military Establishment* (Beverly Hills, Calif.: Sage Publications, 1971, pp. 261–2).

uates rather than promotion from the ranks. Such a decline in the proportion
of commissioned officers coming from the ranks has already been the experience of European all-volunteer forces.[10] A countervailing factor, however,
may be a stepping up of military programs which offer college educations to
highly motivated enlisted personnel.

Developmental Models of the Emergent Military

Underlying much of social change theory are developmental constructs
which are implicit predictions of an emerging social order (for example, a
classless society, a bureaucratic society, a garrison state). Most simply, developmental constructs are modes of analyses which entail historical reconstruction, trend specification, and most especially, a model of a future state
of affairs toward which actual events are heading.[11] Developmental analysis,
that is, emphasizes the "from here, to there" sequence of present and hypothetical events. Put in a slightly different way, a developmental construct is
an "ideal" or "pure" type placed at some future point by which we may
ascertain and order the emergent reality of contemporary social phenomena.
Models derived from developmental analysis bridge the empirical world of
today and the social forms of the future. It follows that one's reading of current and past reality will vary depending upon which developmental model
is constructed.

Our purpose here is to apply developmental analysis to the emergent form
of the military establishment in American society. Put plainly, what is the
likely shape of the armed forces in the foreseeable future? Initially, two
opposing developmental models are presented, each of which has currency
in theories of military sociology. A third model is then introduced which
both synthesizes and differs from the two previous models. All three models,
however, have in common a reference to a continuum ranging from a military organization highly differentiated from civilian society to a military
system highly convergent with civilian structures.

Concretely, of course, America's military forces have never been either
entirely separate or entirely coterminous with civilian society. But conceiving
of a scale along which the military has been more or less overlapping with
civilian society serves the heuristic purposes of highlighting the everchanging
interphase between the armed forces and American society. It is also in this
way that we can be alerted to emergent trends within the military establish-

10. Erwin Häckel, "Military Manpower and Political Purpose," *Adelphi Papers,*
    no. 72 (London: Institute for Strategic Studies, 1970). This is an excellent
    comparative analysis of recruitment and retention policies in Western military
    systems.
11. Heinz Eulau, "H. D. Lasswell's Developmental Analysis," *Western Political
    Quarterly* 11 (June, 1958): 229–42.

ment, trends that appear to augur a fundamental change in the social organization of the armed forces within the near future.

The convergent-divergent formulation of armed forces and society, however, must account for several levels of variation. One variable centers around the way in which the *membership* of the armed forces is representative of the broader society. A second variation is the degree to which there are *institutional* parallels (or discontinuities) in the social organization of military and civilian structures. Difference in required *skills* between military and civilian occupations are a third aspect. A fourth variable refers to *ideological* (dis)similarities between civilians and military men. Furthermore, internal distinctions within the armed forces cut across each of the preceding variables: differences between officers and enlisted men; differences between services; differences between branches within the services; differences between echelons within branches.

Needless to add, there are formidable problems in ascertaining the meaning-evidence on the degree of convergence or divergence between the armed forces and society.[12] Dealing with this issue, some of the more important findings of previous researchers along with the introduction of new materials are given in the developmental models presented below.

Model I: the Convergent or Civilianized Military. A leitmotiv in studies of the military establishment between the wars in Korea and Vietnam was the growing convergence between military and civilian forms of social organization. In large part this convergence was a consequence of changes induced by sophisticated weapons systems. These new technological advances had ramifications for military organization which were particularly manifest in the officer corps. For weapons development gave rise not just to a need for increased technical proficiency, but also for men trained in managerial and modern decision-making skills. This is to say that the broader trend toward technological complexity and increase in organizational scale which was engendering more rationalized and bureaucratic structures throughout American society was also having profound consequences within the military establishment. In the military as in civilian institutions, such a trend involved changes both in the qualifications and sources of leadership.

These changes in military leadership have been examined in several landmark studies dealing with the cold-war military establishment. Ironically, both sympathetic and hostile observers of the changing military establishment have been in accord that there was a convergence in the managerial skills required in both civilian and military organizations. In a highly critical

12. For a somewhat different formulation of the variables involved in a convergent-divergent model of the armed forces and society, see Albert D. Biderman and Laure M. Sharp, "The Convergence of Military and Civilian Occupational Structures Evidence from Studies of Military Retired Employment," *American Journal of Sociology* 73 (January 1968): 383.

and perceptive appraisal of these trends, C. Wright Mills has described the
"military warlords" as constituent members of the power elite in American
society.[13] Mills highlighted the increasing lateral access of military professionals to top economic and political positions. From a different perspective,
other writers—most notably Gene Lyons and John Masland in *Education
and Military Leadership,* and Samuel Huntington in *The Soldier and the
State*—have argued that the complexities of modern warfare and international politics required new formulations of officer professionalization and
civil-military relations.[14]

The most comprehensive study of American military leadership in the
cold-war era has been *The Professional Soldier* by Morris Janowitz.[15] Here,
documentation illustrated the broadening of the social origins of officers to
include a more representative sampling of America's regions and religious
groups, and the increase in the number of nonacademy graduates at the
highest levels of the military establishment. Moreover, the military of that
period was seen as increasingly sharing the characteristics typical of any
large-scale bureaucracy. In effect Janowitz stated that the military was characterized by a trend away from authority based on "domination" toward a
managerial philosophy placing greater stress on persuasion and individual
initiative.

The trend toward convergence has in some respects become even more
pronounced in the early 1970s. Significantly differing from the pre-Vietnam
military, where convergence was most pronounced at military elite levels,
the more recent changes were largely focused—with the accompaniment
of much mass media coverage—on the enlisted ranks. Partly as a result of
internal disciplinary problems occurring toward the end of the Vietnam War,
partly in anticipation of an all-volunteer force after Vietnam, the military
command inaugurated a series of programs designed to accommodate civilian

13. C. Wright Mills, *The Power Elite* (New York: Oxford University Press,
    1956). Similar analyses are found in Fred Cook, *The Warfare State* (New
    York: Macmillan, 1962); and Tristram Coffin, *The Armed Society* (Baltimore:
    Penguin Books, 1964). See also the more recent: John Kenneth Galbraith,
    How to Control the Military (New York: Signet, 1969); Sidney Lens, *The
    Military-Industrial Complex* (Philadelphia: Pilgrim Press, 1970); and Seymour
    Melman, *Pentagon Capitalism* (New York: McGraw-Hill, 1970).
14. Samuel P. Huntington, *The Soldier and the States* (Cambridge, Mass.: Harvard
    University Press, 1957); and Gene M. Lyons and John W. Masland, *Educa-
    tion and Military Leadership* (Princeton, N.J.: Princeton University Press,
    1959). For more recent statements on changing military roles, see Ritchie P.
    Lowry, "To Arms: Changing Military Roles and the Military-Industrial Com-
    plex," *Social Problems* 18 (Summer 1970): 3–16; Robert G. Gard, Jr., "The
    Military and American Society," *Foreign Affairs* 49 (July 1971): 698–710;
    and Sam C. Sarkesian, "Political Soldiers: Perspectives on Professionalism in
    the U.S. Military," paper presented at the annual meeting of the American
    Political Science Association, Los Angeles, 1970.
15. Morris Janowitz, *The Professional Soldier* (New York: Free Press, 1960).

youth values and to make the authority structure more responsive to enlisted needs.

Starting in late 1969, VOLAR (an acronym for Volunteer Army) programs were instituted on a growing number of army posts.[16] VOLAR reforms included such changes as greater margin in hair styles, abolishment of reveille, minimal personal inspections, and more privacy in the barracks. Much of the changed army outlook was captured in its new recruiting slogan, "Today's Army Wants to Join You." The "Z-grams" of Chief of Naval Operations Admiral Zumwalt similarly alerted commissioned and petty officers to concern themselves with enlisted wants and to show more latitude in dealing with the personal life-styles of sailors. The air force, which has always been the most civilianized of the armed services, issued a new regulation in July 1971 (AFR 30-1) which specified a broad set of standards (ranging from haircuts to political protest) which was unprecedented for being equally applicable to both officers and airmen.

Whether changes such as these are really fundamental or are merely cosmetic will take time to tell. But there does seem to be occurring something more than just "beer-in-the-barracks" innovations. Human relations councils consisting of black and white servicemen are coming to play an increasing role in the military's attempt to cope with racial strife. Even more novel are the officially sanctioned councils of junior officers and enlisted men which now exist on a certain number of bases. The formal purpose of such councils is to serve as communication channels between the ranks and the command structure. But a precedent has been established which could be an omen of a major reordering of the traditional chain-of-command authority structure.

Perhaps the sine qua non of a civilian labor force in advanced industrialized societies is collective bargaining of workers. Although trade unionism is hardly more than a cloud on the horizon, there are indirect signs that such an eventuality may someday come to pass. The growing labor militancy of heretofore quiescent public employees at municipal, state, and federal levels may be a precursor of like activity within a future military. Already union membership and military careerism have proved compatible in the military establishments of several Western European countries, notably Germany and Sweden. Even in the United States a precedent of sorts is the situation of full-time National Guardsmen assigned to antimissile installations who are members of state-employee unions. There is also the Trotskyist-influenced American Serviceman's Union (ASU), founded in 1967.[17] In 1971 the ASU claimed an enlisted membership of ten thousand with representation on all major military posts. Extremely ideological and violently hostile toward ca-

16. U.S. Department of the Army, *Project Volunteer in Defense of the Nation,* Executive Summary (Washington, D.C.: Office, Deputy Chief of Staff for Personnel, 1969).
17. An account of the founding of the ASU by its chairman is: Andy Stapp, *Up Against the Brass* (New York: Simon and Schuster, 1970).

reer servicemen ("off the lifers" is an ASU slogan), the ASU's viability as a
genuine trade union is beset by internal contradictions. Nevertheless, the very
existence of an organization such as the ASU is indicative of the incipient
potential for unionization of the military.

At the professional level, the trend toward civilianization is even more
apparent. Among active-duty doctors and lawyers there are manifold indica-
tions of greater identity with civilian professional standards than with those
consonant with military values. (That the chaplain corps seems less likely to
use their civilian clergical counterparts as a reference group is worthy of
note, however.) Most notable, at the service academies the long-term has
definitely been away from traditional military instruction and toward civilianization of both student bodies and faculty—for example, less hazing, reduction of military discipline, more "academic" courses, and civilian professionalization of the teaching staff.[18]

In brief, there is ample evidence to support the model of the military moving toward convergence with the structures and values of civilian society.[19]
This developmental model anticipates a military establishment which will be
sharply different from the traditional armed forces. An all-volunteer membership will be attracted to the services largely on the grounds of monetary
inducements and work selection in the pattern now found in the civilian
marketplace. Some form of democratization of the armed forces will occur
and life-styles of military personnel will basically be that of like civilian
groups. The military mystique will diminish as the armed services come to
resemble other large-scale bureaucracies. The model of the convergent military foresees the culmination of a civilizing trend that began at least as
early as the Second World War and that has been given added impetus by
the domestic turbulence of the Vietnam era.

Model II: the Divergent or Traditional Military. The conceptual antithesis
of the convergent-military model is the developmental construct which emphasizes the increasing differentiation between military and civilian social
organization in American society.[20] Although the consequences of the military buildup caused by the war in Vietnam somewhat obscure the issue, per-
suasive evidence can be presented that the generation-long institutional convergence of the armed forces and American society has begun to reverse itself. It appears highly likely that the military in the post-Vietnam era will

18. Laurence I. Radway, "Recent Trends at American Service Academies," in
    Moskos, ed., *Public Opinion,* pp. 3–35.
19. For additional references to the thesis that the military system will increasingly
    converge with civilian society, see Anthony L. Wermuth, *The Impact of
    Changing Values on Military Organization and Personnel* (Waltham, Mass.:
    Westinghouse Electric Corp., Advanced Studies Group, 1970).
20. For elaboration of the view stressing the divergence of the emergent military
    from civilian society, see my earlier *The American Enlisted Man* (New York:
    Russell Sage Foundation, 1970), pp. 166–82. As is apparent in the conclusions
    of this paper, this is a position I have now come to abandon.

markedly diverge along a variety of dimensions from the mainstream of developments in the general society. This emerging apartness of the military will be reflective of society-wide trends as well as indigenous efforts toward institutional autonomy on the part of the armed forces. Some of the more significant indicators of this growing divergence are summarized immediately below.

First, recent evidence shows that, starting around the early 1960s, the long-term trend toward recruitment of the officer corps from a representative sample of the American population has been reversed. Three measures of the narrowing social base of the officer corps in the past decade are: (1) the overproportionate number of newly commissioned officers coming from rural and small-town backgrounds;[21] (2) the pronounced increase in the number of cadets at service academies who come from career military families;[22] and (3) an increasing monopolization of military elite positions by academy graduates.[23]

Second, although the enlisted ranks have always been overrepresentative of working-class youth, the fact remains that the selective service system, directly or indirectly, infused a component of privileged youth into the military's rank and file. The institution of an all-volunteer force will serve to reduce significantly the degree of upper- and middle-class participation in the enlisted ranks. Since the end of World War II, moreover, there has been a discernible and growing discrepancy between the educational levels of officers and enlisted men.[24] (The 1968 decision to draft a higher proportion of college graduates to meet the manpower needs of the Vietnam War can be regarded as only a temporary fluctuation in this trend.) Very likely, an all-volunteer enlisted membership coupled with an almost entirely college-educated officer corps will contribute to a more rigid and sharp definition of the castelike distinction between officers and enlisted men within the military organization of the 1970s.

Third, the transformation of the armed forces from a racially segregated institution (through World War II) into an integrated organization (around the time of the Korean War) was an impressive achievement in directed social change. Although the military did not provide a panacea for racial problems, it was remarkably free from racial turmoil from the early 1950s through the middle 1960s.[25] It is also the case that the armed forces—as other areas of American life—are increasingly subject to the new challenges of black separatism as well as to the persistencies of white racism. Interracial embroilments have become more frequent in recent years and will al-

21. Radway.
22. Janowitz, "The Emergent Military."
23. David R. Segal, "Selective Promotion in Officer Cohorts," *The Sociological Quarterly* 8 (Spring 1967): 199–206.
24. Moskos, *American Enlisted Man*, p. 196.
25. Ibid., pp. 108–33.

most certainly continue to plague the military. Nevertheless, whatever the racial turn of events within the military, the very integration of the armed forces can be viewed as a kind of divergence from a quasi-apartheid civilian society. The military establishment, albeit with internal strife, will remain into the indefinite future the most racially integrated institution in American society.

Fourth, the well-known trend toward increasing technical specialization within the military has already reached its maximal point. The end of this trend clearly implies a lessened transferability between military and civilian skills. A careful and detailed analysis of military occupational trends by Harold Wool reveals that the most pronounced shift away from combat and manual labor occupations occurred between 1945 and 1957.[26] Since that time there has been relative stability in the occupational requirements of the armed forces. Moreover, as Wool points out, it is often the technical jobs (e.g., specialized radio operators, warning systems personnel) that are most likely to be automated, thereby indirectly increasing the proportions of combat personnel. The use of civilians in support-type positions can be expected to increase with the advent of an all-volunteer force, again thereby increasing the proportion of traditional military occupations within the regular military organization.

Fifth, there is an indication of an emerging divergence between family patterns of military personnel and civilians. Before the Second World War, the military at the enlisted levels was glaringly indifferent to family needs. In World War II, except for allotment checks, families of servicemen more or less fended for themselves. Starting with the cold war, however, the military began to take steps to deal with some of the practical problems faced by married servicemen. An array of on-post priveleges (e.g., free medical care, PX and commissary privileges, government quarters for married noncoms) were established or expanded to meet the needs of military families. This greater concern for service families on the part of the military became especially evident in the late 1960s. Activities such as the Army's Community Service and the Air Force's Dependents Assistance Program are recent efforts to make available a wide range of services for military families: legal and real estate advice; family counseling; baby-sitting services; employment opportunities for wives; loans of infant furnishings, linen, china; and the like. At the risk of some overstatement, the pre-World War II military might be seen as a total institution encapsulating bachelors, while the post-Vietnam military may well encapsulate the family along with the serviceman husband-father.

The above five factors are only a partial list of indicators supporting the developmental model of a divergent military. Mention can also be made of other parallel indicators. Thus charging the armed forces with welfare

26. Harold Wool, *The Military Specialist* (Baltimore: Johns Hopkins Press, 1968).

and job training programs—along the lines of Project 100,000 and Project Transition—can only lead to greater social distance between officers and the ranks. The continuing downgrading of the National Guard and reserve components implies the final demise of the citizen-soldier concept. The further employment of foreign-national troops under direct American command—such as the South Korean troops who today constitute one-sixth of the "American" Eighth Army—would be a paramount indicator of a military force divergent from civilian society.

Perhaps the ultimate indicator of divergence is in the ideological dimension. There is the widespread mood among career officers and noncoms that the armed forces have been made the convenient scapegoat for the war in Vietnam. The mass media, seaboard intelligentsia, and professors of our leading universities are seen as undermining the honor of military service and fostering dissent within the ranks. Although documentation is elusive, the consequence of this has been a spreading defensive reaction within the military community against the nation's cultural elite.[27]

Suffice it to say, there are convincing indicators that the military is undergoing a fundamental turning inward in its relations to the civilian structures and values of American society. With the arrival of an all-volunteer force, the military will find its enlisted membership more compliant to established procedures and a self-selected officer corps more supportive of traditional forms. Without broadly based civilian representation, the leavening effect of relcalcitrant servicemen—drafted enlisted men and ROTC officers from prestige campuses—will be no more. It appears that while our civilian institutions are heading toward more participative definition and control, the post-Vietnam military will follow a more conventional and authoritarian social organization. This reversion to tradition may well be the paradoxical quality of the "new" military of the 1970s.

Model III: the Segmented or Pluralistic Military. In somewhat dialectical fashion the two contradictory developmental constructs of the civilianized versus traditional military can be incorporated into a third formulation: a model of the emergent military as segmented or pluralistic. Such a pluralistic model of the military establishment accommodates and orders the otherwise opposing set of empirical indicators associated with the civilianized or traditional models. Simply put, the pluralistic military will be both convergent and divergent with civilian society; it will simultaneously display organizational trends which are civilianized and traditional.

It must be stressed, however, that the pluralistic military will not be an alloy of opposing trends, but a compartmentalization of these trends. The

27. For perceptive journalistic accounts of growing military estrangement from civilian political and social attitudes, see Ward Just, *Military Men* (New York, Knopf, 1970); and H. Paul Jeffers and Dick Levitan, *See Parris and Die: Brutality in the U.S. Marines* (New York: Hawthorn, 1971).

pluralistic developmental model, that is, does not forsee a homogeneous military somewhere between the civilianized and traditional poles. Rather, the emergent military will be internally segmented into areas which will be either more convergent or more divergent than the present organization of the armed forces. Such a development already characterizes trends between the services. Thus, while the Air Force continues to move toward civilianization and participative control, the Marine Corps announces that it will uphold traditional training procedures and regimentation of personnel. What will be novel in the emergent military, however, is that developments toward segmentation will increasingiy characterize intra- as well as inter-military organization.

Traditional and divergent features in the military will become most pronounced in combat forces, labor-intensive support units, and perhaps at senior command levels. Those in the traditional military will continue to cultivate the ideals of soldierly honor and the mystique of the armed forces. A predilection toward non-civilian values will result from the self-recruitment of the junior membership reinforced by the dominant conservatism of career officers and noncoms. Once beyond the first tour of duty, personnel turnover will be very low. The social isolation of such a traditional military will be compounded by its composition, which will be overrepresentative of rural and southern regions, of the more deprived groups of American society, and of sons of military fathers.

Contrarily, the civilianized or convergent features in the military system will accelerate where functions deal with clerical administration, education, medical care, logistics, transportation, construction, and other technical tasks. Those with specialized education or training will be attracted to the service in a civilian rather than a military capacity and will gauge military employment in terms of marketplace standards. Terms of employment will increasingly correspond to those of strictly civilian enterprises. Lateral entry into the military system, already the case for professionals, will gradually extend to skilled workers and even menial laborers. Concomitantly, there will be a relaxation of procedures required to leave the military. The social composition of such a civilianized military will resemble that of those performing equivalent roles in the larger economy. In all likelihood the present less than two percent female in the armed forces will increase substantially.

From an institutional standpoint, the segmented or pluralistic military will require new organizational forms. The range of such alternative forms can only be sketched here. But as a minimal requirement, there must be some structure which will embrace variegated personnel policies, diverse systems of military justice and discipline, and differing work ethos. Indeed, the antinomies between the civilianized and traditional conceptions of the military may be so great as to prohibit a conventional armed forces establishment. There may develop "two militaries," each organized along entirely different premises. In this format the civilianized military might come to encompass a host of nonmilitary goals—for example, job training, restoring ecologically devastated resources, performing health care services.[28] Another possible alternative may follow the Canadian pattern, where armed forces unification has resulted in a complete separation of support and administrative functions from the combat arms (now referred to as "land, sea, and air environments" in Canadian nomenclature).

Our task here, however, is not to forecast the precise shape of the pluralistic military, but rather to define the constants which will determine the emergent military establishment. Most likely, the armed forces of the United States will keep their overall present framework, but bifurcate internally along civilianized and traditional lines.[29] The traditional or divergent sector will stress customary modes of military organization. In the case of the Army this could entail a revival of the old regimental system. At the same time there will be a convergent sector which operates on principles common to civil administration and corporate structures. Contemporary examples of such organization are metropolitan police forces, the Army Corps of Engineers, and the Coast Guard.

The Emergent Military and American Society

Developmental analysis serves to steer the social researcher between the Charybdis of unordered data and the Scylla of unsubstantiated conjecture as to future social reality. It was with this purpose that three alternative developmental constructs of the military were presented: civilianized, traditional, and pluralistic. And it was the pluralistic or segmented model which seemed to correspond most closely with contemporary trends in emergent military organization.

Ultimately, the implications of each of these models must be assessed for the civil polity and the internal viability of the armed forces. A predominantly civilianized military could easily lose that élan so necessary for the functioning of a military organization. A military force uniformly moving toward more recognition of individual rights and less rigidity in social control would in all likelihood seriously disaffect career personnel while making military service only marginally more palatable to its resistant members. A predominantly traditional military, on the other hand, would most likely be incapable of either maintaining the organization at its required complexity,

---

28. Such a role expansion of the armed services into nonmilitary endeavors is outlined in Albert D. Biderman, "Transforming Military Forces for Broad National Service," paper presented at the Russell Sage Foundation conference on "Youth and National Service," New York, March 1971.
29. In August 1971, the Department of Army announced that soldiers will henceforth be unable to "hopscotch" across military occupational specialties. *Army Times*, 11 August 1971, p. 4. Policies such as these are direct indicators of the move toward a more segmented military.

or attracting the kind of membership necessary for effective performance. More ominous, a traditional military in a rapidly changing society could develop anticivilian values, tearing the basic fabric of democratic ideology.

It is the pluralistic model—with its compartmentalized segments—of the military which seems to offer the best promise of an armed force which will maintain organizational effectiveness along with being consonant in the main with civilian values. Indeed, the model of an emergent military with intra-institutional pluralism may have broader applicability to the framework of the larger social system. Our American society seems to be moving toward a future which is neither a rigid maintenance of the old order, nor an all-encompassing bureaucracy, nor a "greening" of the country. Rather, new forms of voluntarism and counterculture will coexist with persisting large-scale organizations and established values. In the last analysis, the developmental model of a kind of split-level pluralism may well be the defining quality of the emergent American society.

PART III

Africa

THE AFRICAN MILITARY: MODERNIZING PATRIOTS OR PREDATORY MILITARISTS?

Claude E. Welch, Jr.

For most of its history, the United States government has looked askance at military regimes. The conflict-ridden history of Latin America, contrasted with the relatively slight historical influence of the armed forces in the domestic politics of the United States, has helped buttress this opinion. While Bolivia suffered 179 military coups between its independence and 1952, the United States basked in a period of economic prosperity and political calm. Latin America became the continent of the institutionalized coup d'état; North America remained peaceful. While coercion became the main "currency" of political bargaining south of the Rio Grande, or so it was felt in Washington, consent remained the fundamental basis of American politics. The result was a view of the armed forces as "predatory militarists," far more concerned with protection of their narrow self-interests than with overall development of their societies, and voracious for political power but singularly inept in exercising it.

Just over a decade ago, the United States government and many American academics took an extraordinary step, given this historical heritage and set of presumptions. They argued that the armed forces of new states constituted the most satisfactory vehicles for modernization, including long-term development toward democratic practices. This perspective was reflected in the so-called Draper Report, issued by the federal government in 1959, and the same year in an influential article by Guy Pauker, who argued,

Unless drastic changes take place, Southeast Asia will be lost to the free world within the next decade. . . . Communism is bound to win in Southeast Asia . . . unless effective countervailing power is found in some groups who have sufficient organizational strength, goal-direction, leadership, and discipline to be able to compete effectively with the Communists. : . . those best equipped to become an effective counterbalance to the spread of Communism in Southeast Asia are members of the national officer corps as individuals and the national armies as organizational structures. . . . ways must be found to utilize the organizational strength of the national armies and the leadership potential of their officer corps as temporary kernels of national integration, around which the constructive forces of the various societies could rally, during a short period of breakthrough from present stagnation into a genuine developmental takeoff.[1]

Taken together, the Draper Report and Pauker's article presented rosy views of the armed forces as "modernizing patriots." Coups should not be condemned automatically, for the resultant army-based governments might supply the order and organization needed for modernization. The rhetoric of the cold war, admittedly, entered heavily into rationalizing American support for military regimes. But this support was based upon a series of sophisticated arguments. Scholars believed the armed forces possessed a variety of strengths that, they argued, could be transplanted to civilian politics. The military was marked by an efficient, effective, and logical organization; this organization would have a "spillover" effect upon other parts of society. The discipline, hierarchy, and spit and polish of the military should be aped by civilians. Officers did not reflect a narrow, self-serving stratum of society, so the argument went; rather, officers were drawn from the middle or lower-middle class, and would support their interests. As Manfred Halpern argued with regard to the Middle East, widened officer recruitment meant the army has been transformed "from an instrument of repression in its own interests or that of kings into the vanguard of nationalism and social reform." [2] The interests of the military could be directed toward rural development, thus staving off potential insurrection. Expanded responsibilities in "civic action" would enable the armed forces simultaneously to carry out effective counterinsurgency and to give their modernizing proclivities full play. Much of the rhetoric of the Alliance for Progress, and the rationale for extensive American military assistance to Latin America, derived from the hope Castroite uprisings could be precluded by intimately involving the armed forces with social change.

It should be noted that Southeast Asia, the Middle East, and Latin America provided the foci for the "military as modernizers" argument. In the debate that filled hundreds of pages in the early 1960s, surprisingly little attention

1. Guy Pauker, "Southeast Asia as a Problem Area in the Next Decade," *World Politics* 11, no. 3 (April 1959): 337–42.
2. Manfred Halpern, *The Politics of Social Change in the Middle East and North Africa* (Princeton: Princeton University Press, 1963), p. 253.

was given to the armies of tropical Africa by the United States government and American academicians. This oversight in part reflected the continued strong influence of European colonial powers. In 1960, Belgium, France, Great Britain, and Portugal controlled several million square miles between the Sahara and the Zambesi. Even when independence was granted, expatriates continued to dominate the officer corps—the exception being, of course, the Congo, where the mutiny of the Force Publique resulted in the immediate dismissal of all Belgian officers. Military assistance continued to flow from the former colonial powers; the training of African officers continued unabated at Saint-Cyr and Sandhurst. The relatively small amount of United States military aid to Africa reflected this lack of American involvement. Of the total of $33.3 billion in military assistance disbursed by the United States between 1950 and 1968, about one-half of one percent—to be precise, $218 million—went to African states.[3]

The lack of American attention to African armies may also have reflected their small size and political invisibility at the time of independence. The primary function of troops in colonial Africa was suppression of domestic dissent—a task requiring neither extensive training nor equipment, nor even a large number of soldiers. Even as they emerged from the chrysalis of colonialism, the armed forces of tropical Africa remained small: in late 1965, Upper Volta, with an estimated population close to five million, counted an army of 1,500; Nigeria, with a population of over fifty million, counted an army of 11,500.[4] During the period of nationalist pressure for independence, the armed forces stayed politically mute. African states gained self-government by constitutional negotiation, not force of arms. Until the colonial power formally handed over control, the military remained under direct European aegis. Hence the leaders of newly independent African states found themselves occupying presidential palaces, blissfully ignorant of the political potential of their armies.

This ignorance did not last long. A wave of coups started in September 1965.[5] The one-time nonpolitical army, awaiting its orders in the barracks, left the quiet of the encampment for the rough-and-tumble of the political arena. Once members of the military had proved that many vaunted single-party systems had deteriorated into self-serving political machines lacking

3. Ernest W. Lefever, *Spear and Scepter: Army, Police, and Politics in Tropical Africa* (Washington: Brookings Institution, 1970), p. 205.
4. David Wood, "The Armed Forces of African States," Adelphi Papers, no. 27 (London: Institute for Strategic Studies, April 1966), pp. 12, 16.
5. There had been earlier instances of intervention: the assassination of Sylvanus Olympio in Togo; the unceremonious ouster of Maga in Dahomey; Mobutu's 1960 neutralization of Congolese politicians; the East African mutinies of January 1964. However, starting with the 1965 coup in the Congo, intervention became the fad in far more profound ways than the changes dating from 1960 to mid-1965.

popular support, the wave of coups could not be halted. At the start of the 1960s, only Egypt and the Sudan had army-based regimes; by the end of the decade, eighteen African states had experienced successful coups, while many others (notably Gabon, Morocco, and Tanzania) were scarred by putsches or mutinies that had nearly toppled their governments.

The armed forces now serve as the supreme political arbiters in a majority of African states. Coups serve, far more than elections, as barometers of public unrest and as means of political change. Military expenditures weigh ever more heavily upon national budgets. Chou En-lai suggested in 1964 that Africa was "ripe" for revolution. He was wrong: The continent is "ripe" for military intervention, not for popular uprisings. The experience of army rule is one that few countries south of the Sahara will escape.

The first lesson that can be drawn from the military's political success is that civilian governments can be removed without much risk or effort. Rumble a few armored cars into the main squares, incarcerate leading politicians, broadcast appeals for calm, castigate the extravagance and un-representativeness of the ancien régime—these constitute the scenario for most coups d'état. Open fighting has proved the exception rather than the rule.

Why is success so easy? It is not due mainly to high sophistication in coup-making, but to the weaknesses of the governments being deposed. Their basis of popular support has been shown to be remarkably fragile. Few political leaders have successfully counteracted army insurgents by mobilizing urban workers, paramilitary units, or party cadres. The threat of the former colonial power stepping in to reverse a coup has become an increasingly remote possibility since French intervention in Gabon in 1964. The concentration of official functions in the capital simplifies the tasks of coup-makers. The countries of contemporary tropical Africa thus fulfill the three conditions set down by Luttwak for successful intervention: political participation confined to a small segment of the populace; substantial independence and limitation of foreign powers' influence in internal politics; and the existence of a political center, control over which carries with it control of the entire official apparatus.[6]

Having seized power, members of the armed forces must exercise it. Given the relatively small size of the military establishment, a policy common to most African military regimes has been a search for coalition partners. Looking specifically at Ghana and Nigeria, Feit suggested that ruling juntas seek allies among traditional chiefs and bureaucrats.[7] The "outs" of the former government may attempt to regain influence by collaboration

6. Edward Luttwak, *Coup d'Etat, A Practical Handbook* (Greenwich: Fawcett, 1969), pp. 18–46.
7. Edward Feit, "Military Coups and Political Development: Some Lessons from Ghana and Nigeria," *World Politics* 21, no. 2 (January 1968).

with the new regime. The key point is that officers can dispense political power; the longer they enjoy the fruits of control, the more difficult their disengagement from politics becomes.

A third general observation about military regimes in tropical Africa concerns their manipulative view of power. Ruling officers do not wish to taint themselves with open, partisan dispute. Their implicit model of governance is one of control from above; theirs is the "politics of wanting to stand above politics." Yet an inescapable contradiction exists in this view. How can the armed forces both stand above politics, and simultaneously intervene? The usual way out of this dilemma is a stress upon the short-term, cleansing effect of their intervention. Once corruption has been weeded out and other political inadequacies resolved, military leaders promise they will return to the barracks. But this intention has, as yet, rarely been carried out in tropical Africa. Dahomey and Ghana have witnessed the military's voluntary withdrawal from politics—and in both these cases the civilian successor government lasted little more than two years before falling victim to a coup. In other words, by alleging they abhore politics, members of ruling juntas deny the free play of contending factions—and make voluntary withdrawal far less likely than factionalization of the armed forces and further coups d'état.[8] The old adage about mounting the tiger and not being able to dismount applies to those who intervene.

The military's entrenchment in power may be facilitated by the opportunities for self- and corporate enrichment. States in which intervention has occurred, or in which the armed forces exercise considerable political influence, devote markedly higher portions of their budgets to military salaries, perquisites, and equipment.[9] Some analysts have seen intervention as a way for the armed forces to maintain economic privileges.[10] Expansion of the military's political influence hence is linked with expansion of the military's economic demands.

Finally, the army-based regimes of tropical Africa differ markedly. They should not be airily tossed into the same, catch-all category. Their political environments exhibit significant contrasts: In some states, political awareness and participation remain minimal, while other states manifest rapid urbanization and social differentiation. These variations affect the style of military regimes, which I have elsewhere characterized as predatory, re-

---

8. Claude E. Welch, Jr., "Cincinnatus in Africa: The Possibility of Military Withdrawal from Politics," in Michael F. Lofchie, ed., *The State of the Nations: Constraints on Development in Independent Africa* (Berkeley: University of California Press, 1971), pp. 215–37.
9. Eric Nordlinger, *Soldiers in Mufti: The Impact of Military Rule Upon Economic and Social Change in the Non-Western States, American Political Science Review* 64, no. 4 (December 1970): 1135.
10. S. E. Finer, *The Man on Horseback: The Role of the Military in Politics* (New York: Praeger, 1962), pp. 47–56.

form, radical, and guardian.[11] The nature and extent of political participation, to which the conclusion of this chapter will turn, helps answer the predatory militarist or modernizing patriot question.

In order to substantiate these general observations, I shall examine two African states in detail. The armed forces have occupied the major political roles in Egypt for over twenty years, in Nigeria for nearly eight years. Both countries are large: Nigeria counts the largest population of any African state, with fifty-six million inhabitants; Egypt comes second with thirty-four million inhabitants. Both military regimes have attempted major transformations of their political systems, Egypt through land redistribution, widespread economic nationalization, and the would-be development of political movements, Nigeria through dismantling a federal structure that had proved inefficient, corrupt, and illegitimate. The military regimes seem poised to continue in control. Despite occasional, muted references to a possible return to civilian life, the governing juntas of both states will likely remain in control for several more years. The political dynamics unleashed by intervention make simple restoration of a fully civilian government practically impossible. The decision to revise the basic structures of the political system represents a decision that cannot readily be undone.

The coups in Egypt and Nigeria reflected the weak political legitimacy of both governments. In Egypt, the disaster of the 1948 war against Israel, the scandals of arms procurement, the venality of the monarchy, and the government's inability to maintain public order (exemplified in the January 1952 burning of Cairo) testified to political bankruptcy. In Nigeria, political leaders in the Eastern and Western Regions felt increasingly threatened by the growing dominance of the Northern Region in domestic affairs. Blatant election rigging, domestic violence, widespread corruption, and fears of ethnic subordination helped prompt two coups, the first in January 1966, the second six months later. At the time of intervention, accordingly, both states were overripe for change. Members of the armed forces removed the ancien régime by violence. There may have been no alternative.

The armed forces' seizure of power in Egypt and Nigeria was achieved by the actions of a few, closely-linked, lower level officers. Intervention did not result from widespread involvement of army members in politics, but from the conspiratorial efforts of a self-selected handful of officers. The Free Officers in Egypt, and a few Ibo officers in Nigeria, exercised sufficient leverage to bring both countries' governments down like the vaunted house of cards.

The cohesiveness of the Free Officers' inner core has few parallels. Eight of the eleven officers intensely involved in the overthrow of King Farouk were classmates at the Egyptian military academy in the 1936–37 period.

11. Claude E. Welch, Jr., and Arthur K. Smith, *Military Role and Rule* (North Scituate: Duxbury, 1973).

Significantly, the 1936 entering class represented the first opportunity for sons of middle-class Egyptians to qualify for the officer corps. Gamal Abdel Nasser (despite the subsequent protestations of Gen. Mohammed Neguib)[12] was clearly the leader. Nasser personally embodied intense political interests. At one time or another, he belonged to most major political movements in Egypt: the proto-fascist Misr al-fatah ("Young Egypt"), the Muslim Brethren, and the Wafd; at the age of seventeen, Nasser participated in an anti-British demonstration, and for the rest of his life bore the scar of a British bullet in his forehead. Members of the core worked closely in the Palestine War, which transformed their inchoate concerns into detailed plans for intervention. The final plans for the coup were quickly made: Preparation started at 4 P.M., July 22, 1952, and success was assured early the next morning. The Free Officers had the advantage, according to Be'eri, "of being able to act quickly and to muster a powerful force concentrated in a few hands."[13] And, it should be noted, the Free Officers had been a clandestine group, without open ties to political movements. This situation provided them great freedom of action when they assumed control.

Analogous bonds united the conspirators in Nigeria. The seven chief planners (six of whom held the rank of major) had worked near one another during many portions of their professional careers. Three had studied concurrently at Sandhurst; many had shared military postings. This peer group solidarity was reinforced by ethnic ties, for six of the prime movers of intervention were Ibos. Perhaps the dominance of this group should have been expected, given the large portion of junior officers drawn from the Ibos, and given the greater ease with which violence could be plotted among men who shared the same indigenous language and cultural presumptions. Their seizure of control, as noted above, came after a period of growing violence based upon regional and ethnic tensions. The young majors feared the armed forces would be called upon to suppress domestic opposition to the increasingly Northern-dominated government. Faced with the spectacle of a political system ever further removed from popular support, they saw in the arrest and execution of political leaders a simple solution to Nigeria's ills. The federal prime minister, the premiers of the Northern and Western Regions, and the federal minister of finance were among the first victims of military intervention.

As might be expected, the political objectives of the newly triumphant Egyptian and Nigerian officers were obscure and broadly couched. Intervention resulted from reaction against previous excesses. The political order the officers wished to establish was cloudily perceived, at best. They lacked

12. Mohammed Neguib, *Egypt's Destiny* (Garden City, N.Y.: Doubleday, 1955).
13. Eliezer Be'eri, *Army Officers in Arab Politics and Society* (New York: Praeger, 1970), p. 91.

ideologies, though they sounded common themes; their vague ideological preferences led to confusion rather than clarity in decision-making.

The public program of the Free Officers included two planks: "To cleanse the nation of tyrants" and "To restore constitutional life." King Farouk was sent into exile, and an essentially civilian cabinet was installed. The primary political objective of the ruling junta—at this point, still somewhat veiled from public view—appeared to be elimination of contending power groups. Three weeks after the coup, the army crushed a labor uprising; in January 1953, less than six months after the coup, all political parties save the Muslim Brethren were dissolved. The regime coupled a pious pronouncement about a three-year "transitional period" under military dictatorship to the establishment of the so-called Liberation Rally in January 1953; the government vaguely pledged to liquidate imperialism, end feudalism and capitalism, and establish social justice and democratic life. As Dekmejian has caustically commented, "From the outset, the officers' behavior as rulers testified to their total lack of a program for concrete political actions. . . . The haphazard measures of the military at the outset were token reforms and not the implementation of a comprehensive creed of action." [14]

A similar vacuum of ideas marked the army-based regimes of Gen. J. T. Aguiyi-Ironsi and Lt. Col. Yakubu Gowon in Nigeria. Intervention was rationalized as a necessary corrective to previous civilian abuses. The military's rule purportedly would be short, and would be sweet. According to an official pronouncement,

The military government is not an elected government and must not be treated as such. It is a *corrective* government designed to remove the abuses of the old regime and to create a healthy community for a return to civilian government.[15]

But bumbling, fumbling, and lack of follow-through marked the military government's undertaking. It issued regular pronouncements, many of them draconian. Let me cite one notorious example:

Spying, harmful or injurious publications and broadcasts of troop movements or action will be punished by any suitable sentence deemed fit by the military commander. Shouting of slogans, loitering and rowdy behaviour will be rectified by any sentence. . . . Wavering or sitting on the fence and failing to declare open loyalty with the Revolution will be regarded as an act of

14. R. Hrair Dekmejian, *Egypt Under Nasir: A Study in Political Dynamics* (Albany: State University of New York Press, 1971), pp. 23, 50–51.
15. Quoted in Robin Luckham, *The Nigerian Military: A Sociological Analysis of Authority and Revolt 1960–67* (Cambridge, Eng.: Cambridge University Press, 1971), p. 281.

hostility punishable by any sentence deemed suitable by the local military commander.[16]

Rhetoric, not realization, marked the first military government in Nigeria. It preferred the simple, direct word—and penalty—to coalition-building. Concern for national unity paradoxically led them to distrust politicians who might have assisted their efforts. Members of the armed forces apparently felt they could resolve national problems by consulting among themselves. Their policy proclivities could not be readily translated into detailed actions. How is it possible to derive coherent policies from a vaguely described set of preferences that included reduction of disorder, elimination of corruption, opposition to favoritism and nepotism, strong attacks upon regionalism and tribalism, and a high value upon consensus as necessary for the total society?[17] In short, the Nigerian military extrapolated values held dear by the armed forces to the society as a whole.

What accounts for the apparent vacuum of ideas with which the Free Officers and the Nigerian majors undertook to run their respective governments? In Egypt, opposition to the monarchy and its policies did not provide a clear guide to action. The predilections of the Free Officers were veiled by their initial obscurity, first behind the scenes while an essentially civilian government provided a facade, later more prominently behind General Naguib. Nasser and younger officers distrusted Naguib, who seemed more disposed to appeal directly to the people than to those who had guided the coup d'état. In March 1954, Nasser threw down the gauntlet. With his behind-the-scenes support, the Revolutionary Command Council promised an end to military rule by July, including the restoration of political parties and free elections for a constituent assembly. Naguib was caught in a dilemma. He needed a civilian constituency to survive, for his army constituency was threatened by any moves he might take toward restoration of civilian rule. But it quickly became apparent that he lacked sufficiently broad popular appeal. Mobs swarmed through the streets of Cairo, following the announcement that civilian control would return; the throngs demanded that the armed forces remain in command. Naguib had lost his military constituency, and could not build sufficient civilian support; his arrest the following year was a foregone conclusion. With Naguib's elimination, Nasser came into full prominence and started to evolve his political priorities. The major thrust, it became apparent, was the espousal of "Arab Socialism," carrying with it an emphasis on the breakup of large estates, unity among Arab states, and appeals for mass support.

The lack of ideas that marked the Ironsi regime in Nigeria can be attributed primarily to officers' absence of political experience and to the

16. Ibid., p. 292.
17. Ibid., pp. 279–97.

relatively confined circle within which policies were hammered out. The absence of political savvy was not accidental; it had been consciously espoused by the British, who retained substantial direct influence over the Nigerian army until the last expatriate commander departed early in 1965. Anything that smacked of politics was to be avoided. Members of the armed forces had remained neutral in the extraordinary political crises that racked Nigeria in the early 1960s, on the grounds neutrality alone would retain appropriate professional standards and unity. As noted earlier, fear lest the armed forces be called in to prop up the Northern-dominated government helped to prompt the January 1966 intervention.

The life of the regime of General Ironsi was considerably shortened by policy choices he made upon the urging of a small group, predominantly Ibos, gathered around him. In essence, he mistook the support that initially greeted the military's seizure of power with deep-seated support for fundamental reform—notably for the elimination of tribalism and regionalism through increased centralization. The landmark decision was the proclamation of Decree No. 34, which abolished the existing regions, vested all legislative and executive power in the National Military Government, and unified the top levels of the civil service. The decree came as a bombshell. It seemed to confirm increasingly widespread notions of an "Ibo plot" to secure all control. One of its immediate results was rioting against Ibos in northern cities, adding to the rancor and distrust that led to the creation of Biafra and to the bloody civil war.

For the initial months of military government in Egypt and Nigeria, the following observations can be offered. Political consciousness was far more marked among junior officers (particularly majors) than among senior officers. Those who had risen to the top of both armies had extensive military experience but few civilian contacts. Efforts by Naguib to achieve an extramilitary basis for himself succeeded only in antagonizing the armed forces. The political objectives of the newly governing officers were shaped by reaction against the serious political shortcomings that inspired the coups. Rather than create a major program of social, economic, and political change, those who intervened felt they could govern by issuing decrees. Only with the passage of several months—marked in Egypt by the inauguration of the Liberation Rally and the dismissal of Naguib, and marked in Nigeria by Gowon's breakup of the Northern Region and his invitation to Chief Awolowo (both actions necessitated by civil war)—did officers evidence some conception of the political problems that confronted them.

In fact, it was only in 1961, nearly a decade after the Free Officers had seized control, that Nasser and confreres issued a clear set of ideological and policy imperatives. Nasser noted that previous efforts at organizing popular participation—namely the Liberation Rally of 1953–56 and the National Union of 1957–60—aroused little enthusiasm or support. He called for profound change, for the transformation of the National Union "into a revolutionary organ in the hands of the national people." [18] The resulting Arab Socialist Union (ASU) burgeoned. Millions of Egyptians joined: to be specific, 4.8 million by 1964. But the military remained obviously in command. After the crushing defeat by Israel in the 1967 Six-Day War, two-thirds of the leading governmental posts were occupied by officers.[19] And as a result of widespread nationalization, the government payroll swelled, more than doubling between 1962 and 1965. The decisive turn finally had been made. Nasser embarked upon a socialist transformation of Egypt, opening opportunities for at least the lower-middle class, if not necessarily for the *fellahin* (peasants). Civilian allies were co-opted. As Be'eri has documented, the Free Officers expanded political participation and awareness far beyond the narrow coterie of the Farouk era—a "revolutionary change," owing to the elimination of the power of the landed oligarchy. Five groups now constitute the ruling elite: officers, technocrats, some entrepreneurs, medium landowners, and skilled workers.[20]

The cost of the armed forces—which total well over two hundred thousand men—weighs heavily on the Egyptian budget. In 1965–66, defense expenditures reached $480 million; by 1960–70, they had mushroomed to an estimated $805 million. A full 3.6 percent of eligible males have served in the military, a proportion second only to Libya on the entire African continent. The state remains on a constant war footing, with an estimated one-eighth of the Gross National Product fueling the military machine.[21]

With the exception of President Anwar Sadat, all members of the Free Officers' nucleus have died or been ousted from government. Given the change in the ruling elite and the efforts at organizing support through the ASU, does Egypt now exhibit a military regime? The most accurate description is one of a military-civilian coalition, in which members of the armed forces select their civilian allies. The ASU performs a largely ceremonial, hortatory role. It is often remarked that the Middle East exhibits neither war nor peace; the current Egyptian government manifests neither outright military nor total civilian control. Members of the armed forces hold senior governmental positions; the armed forces absorb incredible amounts of money; the co-opted allies have reached a comfortable modus vivendi with the officers. This coalition appears likely to persist.

In Nigeria, General Gowon has committed himself to a return to civilian rule—but only if he is convinced the political system will not slip back to the corruption, pettifoggery, and ethnic preference of the First Republic.

---

18. Dekmejian, *Egypt Under Nasir*, p. 62.
19. Ibid., pp. 174–78.
20. Be'eri, *Army Officers*, pp. 429–39.
21. Figures from the annual survey conducted by the U.S. Arms Control and Disarmament Agency.

Gowon has specified six preconditions for the restoration of civilian government: implementation of the 1970–74 development plan; reorganization
of the armed forces; preparation of a new constitution; agreement upon a
system of revenue sharing among the states; completion of a new census;
and (most difficult of all) the "organization of genuinely national political
parties, which could contend in a general election." As of this date, practically no progress has been made toward achieving the six points, although
a census was carried out in November 1973. The Nigerian army (unlike
the Egyptian army, through the successive parties) has not made a serious
attempt to achieve a firm political foundation. No doubt officers are now
debating whether to link arms with would-be politicians, in order to preserve substantial influence when—and if—the promised handover to civilians materializes, or to delay further steps toward a return to the barracks
in the belief that safeguarding corporate interests requires direct participation in politics. This survery of Egypt and Nigeria illustrates a fundamental
point. Once members of the armed forces develop a clear sense of their
political centrality and confront few obstacles to direct exercise of their
power, military regimes spring up. These regimes, in turn, cannot be easily
displaced.

Intervention came about in the two states without clear ideological guidelines. Early months of military rule witnessed ad hoc decision-making—a
far cry from the rationalism and clear sense of priorities that purportedly
accompany the armed forces. Rough pragmatism best describes these flounder-
ings. The policy proclivities of the new military governors were not clear
guidelines, but invitations to confusion and ever deeper involvement. Internal disputes within the armed forces of both Egypt and Nigeria culminated in substantial personnel changes, at least at the top. The tendency
to rely upon exhortations or commands entailed an inability to compromise,
unite, or inspire. An exception, of course, must be made for Nasser, who
aroused adulation to an extent little comprehended in Western countries.
Even granted these shortcomings, however, the ruling juntas confronted no
serious opponents to their continued work. Having set themselves up as
political arbiters, they could decide whether or not to return the barracks.

The armed forces in politics resembles a man struggling in quicksand.
The more he tries to extricate himself, the deeper he is sucked into the mire.
Intervention is usually a one-way street, leading only to ever-growing mili-
tary involvement in economic, political, and social transformation.

The attractions of power also make the military's withdrawal from direct
political involvement difficult. Governance includes opportunities for prestige, wealth, and psychological satisfaction. The armed forces can quickly
develop a self-serving rationale, that they have been singled out for a task
they uniquely can perform. Consider, for example, this apologia from
Nasser, penned in mid-1953:

The state of affairs . . . singled out the army as the force to do the job. The situation demanded the existence of a force set in one cohesive framework, far removed from the conflict between individuals and classes, and drawn from the heart of the people: a force of men able to trust each other, a force with enough material strength at its disposal to guarantee swift and decisive action. These conditions could be met only by the army. . . . It was not the army which defined its role in the events that took place; the opposite is closer to the truth. The events and their ramifications defined the role of the army.[22]

On the basis of these arguments—the tendency toward the military's increasing involvement, inability to achieve a foundation of political legitimacy, and the presumption of a unique set of political responsibilities—it would appear that the armed forces act as predatory militarists, more concerned with establishing their own position than with effecting necessary change. Yet the other side of the coin must be considered. Both Egypt and Nigeria have experienced transformations, the extent of which the predecessor civilian governments could not have achieved. The significant land reform and nationalization of the economy in Egypt required a determination few political parties possessed to press against the landed oligarchy. The creation of the twelve states in Nigeria could have occurred only with the military's assumption of dictatorial powers, given the strongly entrenched position of the Northern Region. These are no mean accomplishments. But does it make their agents modernizing patriots?

The armed forces played constructive roles in transforming political systems based substantially upon the powers of landed oligarchies or regionally circumscribed elites, through possible alliance with the middle and lower-middle classes. However—save for the charisma of Nasser—the armed forces of Egypt and Nigeria could not inspire widespread political participation or evoke (at least in the short run) a strong sense of national integration. Their seizure of control ousted ineffectual governments, and broadened (especially in Egypt) the ruling elite. In these senses, the military may be deemed modernizing patriots.

By contrast, and in conclusion, the armed forces of Egypt and Nigeria will become predatory militarists the longer their rule persists. The more they remain in their self-appointed political roles, the more they will gain vested interests in retaining power. They may preserve an ossified status quo, not make necessary adaptations. Above all, unless significantly challenged, they cannot organize effective, mass-based political movements without altering their basic styles of decision-making. The armed forces in politics can alter administrative superstructures, engage in economic re-

22. Gamal Abdel Nasser, *Egypt's Liberation: The Philosophy of the Revolution* (Washington: Public Affairs Press, 1965), pp. 42–43.

distribution, attempt to fashion new ideologies of political change, and chart different courses in world politics. Out on the hustings, however, members of the armed forces either bark commands or mumble hortatory slogans. They rarely inspire, they rarely create a strong sense of national unity, and they rarely succeed in removing themselves from the political roles into which they have thrust themselves.

CIVILIAN-MILITARY RELATIONS
IN FORMER BELGIAN AFRICA:
THE MILITARY AS A CONTEXTUAL ELITE*

René Lemarchand

Although coups and military regimes, as Ruth First reminds us,[1] have become a growth industry for academics as well as for military men, the results, for both, have been somewhat inconclusive. As yesterday's strong men yield to new cohorts of aspirants, civilian or military, much of the ongoing debate about the political implications of military interventions in African states seems to take place in an empirical vacuum. Not only is there widespread disagreement about the causes of military interventions, but the more fundamental question of how best to assess or predict the political behavior of African armies once intervention has occurred has yet to receive a definitive answer.

Nowhere has the gap between prediction and reality been more painfully apparent than in the case of former Belgian Africa. The massive killings of which Burundi has recently been the scene provide the latest and most tragic evidence of the futility of some of the prognoses ventured by military analysts: despite the small size of its armed forces, few other states have

---

* Revised version of a paper prepared for delivery at the 1972 American Political
Science Association Annual Meeting, Washington, D.C., September 4–9, 1972.
Although our analysis of the role of the army in Rwanda reflects the situation
existing prior to the coup of July 5, 1973, the rise to power of General
Juvenal Habyalimana does not invalidate the substance of our argument. If
anything the coup of July 5, 1973, confirms the developments I had anticipated
at the time the article was being written (see *infra*, "Rwanda: The GNR as
a Citizen Army).

1. Ruth First, *Power in Africa* (New York, 1970), p. 13.

experienced as many military interventions, and fewer still at a comparable cost in human lives. Not only would the smallness of African armies preclude military take-overs,[2] but, according to some, the absence of an indigenous military elite would serve as a further guarantee of continued civilian rule.[3] Other commentators, succumbing to what in retrospect looks like a fit of delirious optimism, have argued that, irrespective of their position in the political system, African armies would nevertheless provide the civic virtues and technological skills required for rapid development.[4] To systematically disprove each of these contentions would be as facile as it would be invidious. Suffice it to note that few if any of the major hypotheses thus far advanced by social scientists provide adequate explanations for the behavior of the military in Zaire, Burundi, or indeed Rwanda.

This unhappy state of affairs is perhaps less a reflection of the presumed uniqueness of the history of Belgian colonial rule or of the societies upon which it was forced as of the difficulties inherent in the assessment of African military regimes in general. Compared to students of the military in Latin America, Africanists are at a distinct disadvantage in gathering and handling their data: the very recency of military rule in Africa makes it as yet impossible to use meaningful time-series in analyzing the performance of military regimes; ethnicity, both within and outside the military, introduces a variable which in the case of Latin America has little or no relevance; more important still, access to such basic data as recruitment patterns, social background characteristics, the ratio of officers and noncoms to enlisted men, etc., is less difficult in Latin America. What further complicates the task of the Africanist is the comparatively low level of professionalization of African armies, a situation which, besides greatly inhibiting access to information, generates a type of civilian-military relationship that is peculiarly difficult to define and investigate, as the contours of the polity are necessarily unstructured and relations among actors exceedingly personalized and fluid. For a Latin American equivalent of these inchoate, machine-like characteristics of military regimes, one must move back in time to the pre-World War II period—if not earlier. If any parallel can be drawn between the Mobutu and Micombero regimes, the analogy that comes to mind is neither Brazil or Argentina in the 1970s, but Cuba and Ecuador in the 1920s and 1930s.[5]

In a more specific sense, one might argue that many of the shortcomings of studies of the military in Africa stem from the same lack of sensitivity to

2. See W. Gutteridge, *Military Institutions and Power in the New States* (New York, 1965), pp. 143–45.
3. E. Shils, "The Military in the Political Development of the New States," in John J. Johnson, ed., *The Role of the Military in the Underdeveloped Countries* (Princeton, 1962), p. 54.
4. See Lucian Pye, *Aspects of Political Development* (Boston, 1966), pp. 172–87.
5. I am indebted to my colleague, Sam Fitch, for suggesting the analogy.

environmental variables that characterized earlier studies of the military in Latin America. In each case the tendency has been to overestimate the explanatory power of military institutional characteristics while giving short shrift to interaction situations between the political system and the military subsystem. What Stepan refers to as the "unitary, self-encapsulated aspects of the military institution" [6] has thus usually been treated as the only significant independent variable for the analysis of military regimes. Far less attention has been paid to the pressures originating from the political system, and how these pressures may in turn condition military responses. Yet the evidence shows that in Stepan's words, "in many developing countries not only is the military not isolated from the tensions experienced by the general populations and therefore not able to act as an integrating force, but the military is itself an element in the polity that may transform latent tensions into overt crises." [7]

The aim of this article is to explore the implications of this hypothesis in the light of empirical evidence derived from Zaire, Burundi, and Rwanda. The main justification for the selection of these three states is that, although they share important similarities in terms of their colonial experience, and in the case of Rwanda and Burundi in terms of their geographical scale, social structure, and traditional political organization, they have nonetheless evolved radically different patterns of civil-military relations. Ironically, those two countries that had most in common prior to independence (Rwanda and Burundi) are now at polar opposites on the continuum of civil-military relations. While Rwanda is still enjoying the blessings of civilian rule, the army in Burundi, or what is left of it, directly controls the destinies of the country. Zaire, on the other hand, can perhaps best be located at a point somewhere between these two extremes: although the polity retains the formal trappings of civilian rule, the army nonetheless acts as the most powerful and influential pressure group in the political system.

In seeking to elucidate these variant patterns of relationships, this analysis employs the following general assumptions. (1) Just as the political crises experienced by each state at the time of independence have shaped the institutional characteristics of their respective armies in specific ways, these characteristics have subsequently conditioned the responses of their military elites to political pressures along different lines. (2) Despite these institutional differences, in each state, the attitude of the military (or a segment thereof) has tended to reflect their changing perceptions of the threats posed to their corporate, ethnic, or class interests by the political environment. (3) In each state environmental threats involved a combination of internal and external threat perceptions; specifically, the spillover of anti-regime forces across

6. A. Stepan, *The Military in Politics: Changing Patterns in Brazil* (Princeton, 1971), p. 10.
7. Ibid., p. 10.

national boundaries has had a direct impact not only on military-civilian relations within the host country but on transnational patterns of interaction, heightening the chances of involvement of each state in the internal affairs of its neighbor.

Our universe of comparison thus encompasses more than just three "national" units of analysis; it also includes different levels of analysis—institutional, national, and international. These are perhaps best seen as intersecting worlds. The penetration of refugee groups from Zaire and Rwanda into the domestic sectors of Burundi, the extension of training facilities by Zaire to the military forces of Burundi, the "demonstration effect" of the Rwanda revolution on the polarization of ethnic feelings in Burundi all bear testimony to this overlap. This qualification is intended both as a corrective to the conventional view that institutional, national, and international spheres merely stand in a relation of greater inclusiveness to each other, and as a logical entry point into the question of how the colonial "inheritance situation" [8] has affected the shape of military formats in each state.

The Inherited Military Formats

In spite of their common colonial experience, the security forces of Zaire, Burundi, and Rwanda were shaped by very different historical factors. Not only were they established at separate points in time, but on the basis of different policy considerations and in radically different cultural and political contexts.

Whereas the origins of the *Armée Nationale Zairienne* (ANZ) are traceable to the *Force Publique* (FP), formally established in 1888, the initial steps that led to the creation of the *Garde Nationale Rwandaise* (GNR) and the *Armée Nationale du Burundi* (ANB) were not taken until 1960, as belated attempt to fill the security void anticipated by the imminent withdrawal of Rwanda and Burundi from the "jurisdiction" of the FP. By 1961 the *Garde Territoriale Rwandaise* (subsequently renamed *Garde Nationale Rwandaise*) numbered 680, and its Burundi counterpart approximately 900. In the same year officer schools were hastily set up in Kigali (Rwanda) and Bujumbura (Burundi), and a small number of cadets sent to the *Ecole Royale Militaire* (ERM) in Belgium.[9] Not until 1963, however, a year after the independence of Rwarda and Burundi, would the first batch of officers be available for a partial Africanization of their command structures. Compared to the FP, then, the security forces of Rwanda and Burundi are evidently very recent creations. Yet, at the time of independence, the pace of

8. The term is borrowed from J. P. Nettl and R. Robertson, *International Systems and the Modernization of Societies* (New York, 1968).
9. For further information, see *Procès-Verbal de la Réunion de Tutelle du 14 mars 1961*, mimeographed (Kigali, 1961).

their Africanization programs was comparatively more advanced than in Zaire, a fact which in retrospect seems hardly surprising given the extreme belatedness and impracticality of the efforts made by Belgium to Africanize the command structure of the FP. Out of a total of 25,000 men, only three had achieved sergeant major rank at the time of independence.

Unlike its counterparts in Rwanda or Burundi, the FP, as has often been emphasized, was essentially designed as an instrument of coercion in the hands of the colonizer. Neither intended to serve as a channel of upward mobility nor as the institutional preserve of "warlike tribes," "it was above all an instrument of repression against potential troubles among the civilian population." [10] Its operational code, as one scholar puts it, was one of "gangsterism." [11] Its position in society was comparatively isolated, and its recruitment patterns based on a careful scrambling of its ethnic components. As the *grande muette* of the Belgian Congo, the FP was expected to stay out of politics in order to live up to its tradition of loyalty to the state. Very different were the motives behind the policies of the administering authority in what was then known as the UN Trust Territory of Ruanda-Urundi. In each of the emergent successor states, security was defined according to the colonizer's own estimate of the political forces at work. The aim was not so much to insure the loyalty of the military to a political abstraction—the state—but to a specific party or group of politicians. This, in effect, meant loyalty to that party or faction which, in the eyes of the administering authorities, claimed political legitimacy.

Although these contrasting conceptions of "security" are helpful in explaining the relative isolation of the FP, as compared with the *Gardes Territoriales* in Rwanda and Burundi, attention must also be paid to the different ethnic and political contexts in which security forces were born. In contrast with the situation prevailing in Zaire, where cultural pluralism takes the form of vertical cleavages among a multiplicity of discrete ethnic units, prior to independence Rwanda and Burundi were highly stratified societies, with power gravitating in the hands of a traditional aristocracy which also claimed the prerogatives of a ruling caste.[12] In each state political and economic cleavages tended to coincide with ethnic divisions. In each state the Tutsi minority stood in a position of virtual supremacy in relation to the Hutu masses. While both societies exhibited very sharp differences with their neighbor to the west in terms of scale, social structure, and political organization, between them the similarities should not be exaggerated. The greater complexity of Burundi society—reflecting the privileged position of the

10. J. C. Willame, *Patrimonialism and Political Change in the Congo* (Ph.D. diss., University of California at Berkeley, 1970), chap. 3, p. 9.
11. Morris Janowitz, *The Military in the Political Development of New Nations* (Chicago, 1964), p. 34.
12. For further information, see R. Lemarchand, *Rwanda and Burundi* (New York and London, 1970).

princely *ganwa* oligarchy in the political system, as well as the presence of significant intracaste cleavages among both Hutu and Tutsi—was instrumental in delaying the occurrence of a Rwanda-like Hutu-Tutsi confrontation, and at the beginning in mitigating the pressures of ethnicity on the army. In Rwanda, by contrast, the Hutu revolution of 1959 resulted in a sharp and abrupt polarization of ethnic feelings even before any concrete step had been taken to set up an army. Although the ethnic problem has since been resolved in different ways in each state, at the time of independence ethnic pressures were certainly higher in Rwanda and Burundi than in Zaire, and higher still in Rwanda than in Burundi. Recruitment patterns, as we shall see, were directly affected by these contextual variations.

The foregoing considerations give us several important clues to an understanding of the institutional characteristics of the armed forces of each state, and of their relationships to their respective political systems. For one thing, the very recency of the process of military institutionalization in Rwanda and Burundi virtually eliminated the chances of armed revolts at the time of independence. At the peak of the Rwanda revolution the army had yet to materialize. When mutinies occurred, they were among the detachments of the FP stationed in Kigali,[13] prompting the local authorities to rely on metropolitan forces for the maintenance of "peace and order."

Moreover, the somewhat more accelerated pace of Africanization in Rwanda and Burundi, while further contributing to minimize the risk of armed mutinies, lessened the chances of an abrupt and massive promotion of noncommissioned grades as happened in Zaire in the wake of the mutiny of the FP. Thus there is no parallel in Rwanda or Burundi for the deep generational cleavage and mutual animosities discernible between the old noncoms of the FP who were promoted overnight to the rank of general or colonel,[14] and the younger generations of officers trained at the ERM. Nor is there any parallel in either state for the professional limitations which this situation imposes on the ANZ high command. Only among the more recently trained and younger generations of officers can one detect an emergent sense of military professionalism; at the senior level the ex-noncoms of the FP continue to display attitudes reflective of their previous training, or lack of training. Thus unlike what can be seen in Rwanda and Burundi, where ethnic

13. Indeed, almost totally unnoticed by outside observers is the mutiny which took place in mid-1960 among the FP detachments stationed in Kigali, apparently motivated by the fear that, following the breakdown of the Congo-based FP, Europeans might seek revenge against the Rwanda-based detachments. After communicating their demands to the Belgian commander, Colonel Logiest (soon to be appointed *Résident Spécial* for Rwanda), the Congolese soldiers were allowed to return to the Congo—with Logiest momentarily held as a hostage by the troops to insure their safe passage across the border. See J. H. Hubert, *La Toussaint Rwandaise et Sa Répression* (Brussels, 1965), p. 52.

14. Crawford Young, *Politics in the Congo* (Princeton, 1965), p. 446.

solidarities are all-pervasive (though not immutable), peer-group solidarities are on the whole far more pronounced within the command structure of the ANZ. Not only does this situation create serious tensions between the ANZ high command (almost exclusively composed of former FP noncoms) and the younger generations of noncommissioned officers, but it also tends to generate rather different networks of relationships between army men and civilians.

In sharp contrast with the policy of ethnic scrambling pursued in the Congo, recruitment policies in Rwanda and Burundi were much more deliberately "selective." The resultant ethnic profiles of their armed forces were by no means identical, however. While in Rwanda every effort was made to incorporate as large a number of Hutu elements as seemed compatible with the policy of "fairness" prescribed by the UN, in the end resulting in an all-Hutu army, recruitment policies in Burundi produced a far more variegated picture, with Hutu and Tutsi elements almost equally represented within the officer corps.[15] The assumption was that the military in order to be politically reliable should reflect the same ethnic balance as the parties then in existence; in the absence of strong revolutionary undercurrents there seemed no reason at first to favor any particular ethnic group, as in Rwanda. In both cases, however, the main criterion of selection was the presumed loyalty of army recruits to the party favored by the Belgian authorities, the *Parti de l'Emancipation de Peuple Hutu* (*Parmehutu*) in Rwanda, and the *Parti Démocrate Chrétien* (PDC) in Burundi.

Unlike the ANB or the FP the GNR was not intended to be a "national" army in the sense of being recruited among the major ethnic segments of society, but rather, like Cromwell's army, was meant to "represent the particular sect and the particular party which had gained the upper hand in the Civil War."[16] The GNR, in short, was intended to serve as the shield of the Hutu revolution. The ANB, by contrast, represented an ethnic "mix" which naturally increased its potential vulnerability to ethnic pressures. As a result the army eventually came to supplant both party and parliament as the main arena of ethnic confrontation, leading to major purges of Hutu officers in 1966 and 1969, and culminating in 1972 with the alleged massacre of some 450 Hutu troops and all remaining Hutu officers.

The military formats inherited by each state have thus had a direct impact not only on the institutional characteristics of their armed forces but also on the relationships of these forces to their respective social systems. Focusing on the patterns of *ethnic congruence* discernible between the military and society, at the time of independence the Rwanda situation differed markedly from that of Zaire and Burundi: in one case a single ethnic group dominates all rank levels in the army; in the other two the balance of ethnic

---

15. See Lemarchand, *Rwanda and Burundi*, p. 461.
16. C. H. Firth, *Cromwell's Army* (London, 1902), p. 346.

interests within the army closely approximates the distribution of ethnic groups in society as a whole. Rather than a product of military profession-alism, continued civilian control of the military in Rwanda is best explained in the light of the dominant position assumed by Hutu elements in both the army and society. Looked at from the perspective of *social stratification,* the case of Burundi differs from both Rwanda and Zaire: rather than repre-senting a reversal of traditional social statuses (as in Rwanda), or the ex-tension within the army of the predominant mode of social stratification (as was the case of the FP under colonial rule), at the time of independence Burundi stood halfway between these two types, incorporating within its armed forces fundamental ethnic and social contradictions. Again one might concentrate on the extent to which military cleavages tended to replicate *political cleavages* in society and then end with yet another type of classifica-tion, depending on the time period considered.

The one common characteristic shared by all three armies was their ex-tremely low level of professionalization. Officer training in the Congo did not commence until after independence, and in the case of Rwanda and Burundi only a year before independence, and then on a very limited scale. As of March 1961, for example, the Kigali officer training school had a total enrollment of seven candidates (one of whom was Tutsi), with only two cadets attending the ERM in Brussels.[17] Although the training program envisaged at the time involved a preliminary eight-month period at the Kigali school, followed up by a twenty-month training period in Belgium, the length of military training abroad was in many cases reduced to six months in order to meet the exigencies of a *formation accélérée.* Thus out of the nineteen officers and noncommissioned officers who sat on Burundi's *Conceil Suprême de la République* in 1972, ten had undergone a six-month training period at the Arlon Infantry School, and only one (a major) had gone to St. Cyr for a three-year training period. One might also note in passing that the quality of the officer training provided at Arlon does not compare very favorably with the training offered at the ERM, or for that matter at St. Cyr or Sandhurst. The least one can say is that the content of the curriculum at Arlon is rather ill-suited to bring about a change of traditional reference-group identifica-tions. The comparatively weak impact of military socialization on their of-ficer corps substantially enhanced the weight of primordial reference-group identifications of the kind noted above. How these have, in turn, conditioned patterns of civil-military relations since independence is what we shall now examine.

Patterns of Civilian-Military Interaction

Since we are primarily concerned in this section with the analysis of interaction situations between civilian and military elites, it is worth noting

17. See *Procès-Verbal de la Réunion de Tutelle, op. cit.*

at the outset that the term "elite" is in each case highly relative. It is relative, first, to community standards. Prior to independence, the *évolués* of the Congo were by local standards an elite, as were, immediately after independence, the sergeants and privates who were overnight catapulted to the rank of colonel. Community standards are by no means fixed, however, and what might have been considered as elite criteria at the time of independence might not be so considered a decade later. In terms of education and training, for example, the ANZ noncoms currently attending the Luluabourg officer school have stronger claims to being an elite than many of their commanding officers.

The term elite, therefore, is also relative in that, as Wilkinson puts it, "high status and sense of group are themselves relative attributes." [18] How much more of an elite is a second lieutenant after a six-month training period at Arlon, compared to a colonel after a three-year stay at the ERM? Or, to take an example from Burundi, how much more of an elite is a Tutsi lieutenant colonel with a six-month training at Arlon, compared to a Hutu major with three years at St. Cyr? Both undoubtedly enjoy high status. But do they feel that they share the same elite characteristics? Their perceptions of themselves as members of an elite group are by no means fixed or uniformly defined. Just as changes in community standards are likely to bring about new patterns of interaction between military and civilian elites, changing perceptions of what constitutes the mark of an elite may have a direct impact on interaction situations within each group.

Elite status is relative, too, in that it can be defined according to different criteria depending on what reference-group identifications are operative at a particular time in a particular setting.[19] In none of the three states under consideration are the self-images of the military definable solely in terms of their corporate membership in an officer corps. We are here dealing with what might be referred to as *contextual elites*,[20] inasmuch as their self-images and self-interests are likely to vary in relation to contextual variables. Ethnic, class, or peer-group interests are by no means fixed quantities, any more than the perceived threats to these interests. Environmental challenges, whether arising from specific policies or unanticipated occurrences, are bound

18. Rupert Wilkinson, ed., *Governing Elites* (New York, 1969), p. xiv.
19. Much of the point regarding the relevance of "reference-group identification" was inspired by Robert Price's seminal article, "A Theoretical Approach to Military Rule in New States: Reference Group Theory and the Ghanaian Case," *World Politics*, XXIII, 3 (1971), pp. 399–430. I am grateful to Sam Fitch for directing my attention to this article.
20. The difference between such an elite and what Stepan refers to as a situational elite lies in the fact that the latter derive "their power and prestige from their membership in an institution with power" (as in the case of the Brazilian military elites described by Stepan). Contextual elites, on the other hand, have no single frame of reference, whether institutional or societal, to define their self-images and group interests. See Stepan, *Military in Politics*, p. 270.

to affect the nature (or saliency) of reference-group identifications (corporate, ethnic, regional) among military elites, as well as the manner in which they perceive environmental threats to their interests.

Although attempts at differentiating between military and civilian elites must necessarily take into account the variability of their respective reference-group identifications, it is nonetheless possible to sketch areas of congruence between them at any given time. Table 1 suggests what these patterns are in each state, as of June 1972.

Table 1: *Patterns of Civilian-Military Congruence in Rwanda, Burundi, and Zaire* (1972)

Reference Group	Strength of Civilian-Military Identification			Nature of Primary Civilian-Military Linkages		
	Rwanda	Burundi	Zaire	Rwanda	Burundi	Zaire
Ethnicity	high	high	mixed	Ethnic	Ethnic	—
Region	low	mixed	mixed	—	—	—
Class	low	high	high	—	Class	Class
Age Group	low	high	low	—	Age	—

The limitations of this type of schematization are all too evident: The variables we have outlined are of little help to an understanding of sequential variations. Nor do they convey a very accurate picture of the individual variations among these states; like all "ideal" representations, they greatly oversimplify political realities. Nor can one expect such a formulation to shed much light on interaction patterns at the level of noncommissioned ranks, as contrasted with the army command or the officer corps. Nonetheless, these variables set the key parameters within which civilian and military elites interact: the higher the degree of ethnic congruence between them, the greater the chances of interaction situations clustering around, say, regional, class, or generational issues; conversely, the higher the degree of "class" congruence, the greater the probability of ethnic or other variables intruding themselves into the sphere of civilian-military relationships. Where "class" and ethnicity rank equally high, however, the economic wolf in tribal sheep's clothing, as Sklar puts it, is admittedly more difficult to identify.[21]

These characterizations, moreover, need to be supplemented by an analysis of the control mechanisms used by civilian elites to retain the support of the military, or vice versa. What follows, then, is an attempt to move beyond the somewhat static formulation suggested in table 1 and look at civilian-military relations in each state in the light of specific action patterns.

21. Richard Sklar, "Contradictions in the Nigerian Political System," *Journal of Modern African Studies* 3:2 (1965).

Zaire: The Praetorian Syndrome

"A modern praetorian state," writes Perlmutter, "is one in which the military tends to intervene and potentially could dominate the political system. The political process of this state favors the development of the military as the core group and the growth of its expectations as a ruling class; its political leadership . . . is chiefly recruited from the military, or from groups sympathetic or at least not antagonistic to the military." [22] Formulated in these terms, praetorianism is clearly the hallmark of the Zairian polity. The army forms the core of the political nucleus around which Mobutu seeks to structure his authority, and the balancing act that holds the state together ultimately depends on his ability to play civilian and military factions off against each other. Although the political system retains the formal trappings of civilian government, the army exercises control behind the scenes and on specific matters. Its role, basically, is that of an empire. In line with Mobutu's alleged predilection for the Kemalist model, the ANZ sees itself as the custodian of civilian authority. Far from acting as "an organism of execution operating in a strictly military context," as Mobutu described it in 1965, the ANZ is vitally concerned with politics, and openly admits it: "For us, soldiers, the Congolese people must confirm their aspirations to peace [by their electoral choice]. . . . We have the right to concern ourselves with [their] choice because we have been the principal architects of restored peace." [23]

The emergence of the ANZ as an arbitrator-type praetorian army reflects the added weight it has gained in the political system since 1960, as well as the still very low level of political institutionalization and support for the political scaffolding built by Mobutu since 1965. The current position of the ANZ in the political system is perhaps best understood in the light of the changes that have taken place in civilian-military relations since independence.[24] Broadly speaking, three major phases can be distinguished in the recent history of the Zairian armed forces: (1) the "rabble" phase, coinciding roughly with the FP mutinies and their aftermath (June–September 1960), in which the character of civilian-military relations can best be summed up by Voltaire's mot about eighteenth-century Prussia: "A country that did not have an army as the army had the country"; (2) the "private militias" phase, lasting from September 1960 to November 1965, in which each of the main contending political figures sought to recruit private armies

---

22. Amos Perlmutter, "The Praetorian State and the Praetorian Army," *Comparative Politics* 1, no. 1 (1969): 383.
23. *La Voix de l'Armée Congolaise,* 1 November 1970.
24. For an illuminating discussion of the state of the military in the Congo during this phase, see Willame, *Patrimonialism and Political Change in the Congo., op. cit.*

among their respective groups of origin, leading to the emergence of four separate armies: (a) the South Kasai Army of Albert Kalondji, (b) the army of the Independent State of the Katanga, loyal to Moise Tshombe, (c) the National Congolese Army of Antoine Gizenga, and (d) the National Congolese Army of Joseph Mobutu; (3) the "praetorian military" phase, from November 1965 to the present, in which patronage serves as the cement that holds the civilian and military elites together, and distrust of first generation politicians as the principal justification for the army's arbitration of civilian politics (see table 2).

Table 2: *Evolving Patterns of Civilian-Military Relations: Zaire, 1960–72*

Time Period	Type of Military Establishment	Recruitment Patterns	Type of C-M R	Control Mechanisms
1. June–Sept. 1960	"Rabble"	Mixed	Unstruc-tured	None
2. Sept. 1960–Nov. 1965	Private Militias	Predomi-nantly "Tribal"		
	1. South Kasai Army	Baluba	Fragmented, Personalized. Shifting Civilian-Military Alignments.	Ethnic Clientage & Personal Loyalty
	2. Katanga Army	Bayeke-Lunda		
	3. Stanleyville Army	Batetela-Bakusu		
	4. Leopoldville Army	Mixed		
3. Nov. 1965–June 1972	Praetorian Military	Mixed (with predomi-nance of Bangala in some units)	Mutual Arbitration	Patronage

Basically the trend has been from a situation where the intrusion of political divisions within the military causes its fragmentation into opposing factions, to one in which the military intervenes into politics in order to neutralize divisions among civilian politicians.

In many ways, of course, the ANZ still bears the markings of its institutional antecedents. Contacts between the troops and the civilian populations are still characterized by considerable brutality and arbitrariness, and evidences of ethnic favoritism in promotions and recruitment patterns are still too numerous not to cast doubts on its "national character." Yet there is no gainsaying the significance of the institutional changes that have taken place since 1965. The result has been, in Willame's terms, "the substitution of a trained army for a blind instrument of repression, the birth of a military intelligentsia, and the absorption of considerable technical resources by the army." [25]

Specifically, there has occurred a substantial increase in the number of promotions to senior ranks between 1965 and 1967: While the number of generals and colonels increased respectively from 3 to 13 and 14 to 56, the number of majors jumped from 55 to 93; concomitantly, however, the total number of middle-rank officers (lieutenants and second lieutenants) dropped from 555 to 380.[26] Second, a determined effort has been made to improve officer training facilities. A number of new officer training schools have been set up at the cadet, company commander, and specialist levels (such as the *Ecole de Formation des Officiers* (EFO) and the *Ecole des Commandants de Compagnie* (ECC), both at the *Centre Militaire de Luluabourg*, and the newly created *Centre Commando de Kota-Kati*, in the Equateur). The aim, presumably, is to produce officers who will combine the qualities of "hommes d'honneur" and "citoyens d'élite." [27] Third, there has been a spectacular improvement in the technical capacity of the ANZ over the last five years, reflecting a concomitant growth of imported military hardware. The net effect of these changes has been to give the ANZ an unprecedented awareness of its coercive potential, and a renewed confidence in its ability to arbitrate. By the same token, however, the rather lopsided character of promotions within the officer corps, together with the changes taking place in the self-images of the noncommissioned officers, are likely to accentuate generational tensions between the top command posts, still occupied by former FP noncoms, and the noncommissioned ranks.

To prevent the army from arbitrating its own divisions and at the same time strengthen its loyalty to the regime involves more than just meeting the professional demands of the officer corps. It also requires the use of strategies designed to solidify its links with society. Of these strategies, at least three deserve mention. First, increasing attention is being paid to ethnic and regional criteria in the recruitment of the armed forces. Thus, while the Equateur contributes more recruits to the ANZ than any other province,[28] the Thysville-based Armored Brigade and the Binza-based Paracommandos are said to be composed almost exclusively of Bangala elements. Inasmuch as it helps strengthen regional and cultural ties between the praetorian president and his "guards," ethnicity is a critical factor in securing the loyalty of the ANZ to the regime. Second, every effort is made to maintain and indeed

25. J. C. Willame, "Congo-Kinshasa," in C. Welch, ed., *Soldier and State in Africa* (Evanston, 1970), p. 137.
26. See B. Verhaegen, "L'Armée Nationale Congolaise," *Etudes Congolaises* 10, no. 4 (1967): 14.
27. *La Voix de l'Armée Congolaise*, 1 November 1970.
28. Verhaegen, "L'Armée Nationale Congolaise."

reinforce functional divisions within the armed forces, so as to maximize the chances of retaining the support of at least some units in case others prove disloyal. In this respect both the Thysville Brigade and the Paracommandos perform a major security function on behalf of the regime. Like others before him, Mobutu, "sensible that laws might color, but that arms alone could maintain his usurped dominion . . . distinguished those favored troops by a double pay and superior privileges; but, as their aspect would at once have alarmed and irritated the . . . people, [two] cohorts only were stationed in [or near] the capital." [29] Third, and most important, a growing proportion of the national budget is devoted to meet the demands for higher pay and perquisites voiced by the officer corps.

The "class" character of the officer corps constitutes one of the most important linkages between military and civilian elites, creating the basis for a relatively stable political alliance between them. This is not meant to suggest that ANZ officers form a dominant class in the Marxist sense. They clearly have no direct participation in the process of production. That they do constitute an economically privileged stratum is nonetheless undeniable. What evidence is available suggests a situation somewhat similar to that described by Lofchie in his discussion of the military in Uganda: "The cumulative impression is of an overbearing military establishment entirely unaccountable to civilian authority, engaged in prodigious misuse of public funds, and involved in immense profiteering at the expense of the society of which it [is] a part." [30]

The evidence, in part, comes from the army itself. In an editorial entitled "La subversion, la corruption, nos deux grands fléaux," the *Bulletin Militaire* of April 1964 publicly admitted the existence of corrupt practices within the military. The cases handled by military courts in 1965 are equally revealing: according to Willame, "During 1965 one Colonel, several Majors, dozens of officers, hundreds of noncommissioned officers and soldiers were sentenced to long-term imprisonment for corruption, lack of discipline, and mismanagement. Several among them were executed by firing squads." [31] As late as 1971, however, several similar cases again came up before military courts.

Excess expenditures over the military budgets for 1965 and 1966 supply additional evidence of the "demands" made by the army. In each year actual expenditure exceeded respectively by 20 and 15 percent the amounts originally allocated to the armed forces. One might also note in this connection the increasing proportion of budgetary expenditures devoted to the army

29. E. Gibbon, *The History of the Decline and Fall of the Roman Empire* (London, n.d.)1: 43.
30. Michael Lofchie, "The Uganda Coup—Class Action by the Military," *Journal of Modern African Studies,* April 1972.
31. J. C. Willame, *Patrimonialism and Political Change in the Congo,* chap. 6, p. 35.

from 1964 to 1969, which jumped from 13 percent of the national budget in 1964 to 21.9 percent in 1967, and from a total of twelve million zaires in 1967 to twenty-seven million in 1969.

Table 3: *Defense Expenditures: Zaire, 1964–69*

Year	Defense Budget (in zaires)	Percentage of Total Defense Expenditures
1964	4,593,000	13 %
1965	10,265,000	18.6%
1966	13,000,000	21.3%
1967	12,792,000	21.9%
1969	27,000,000	14.6%

Source: H. Leclercq, "Evolution des Finances Publiques de 1966 à 1969," *Cahiers Economiques et Sociaux* 7 (1969): 159; and R. Booth, *The Armed Forces of African States* (London: Institute of Strategic Studies, 1970).

Not only does the military as a whole receive a disproportionately large share of the financial pie, but the involvement of ANZ officers in commercial transactions allows them to substantially increase their regular salaries. Especially damning in this respect is the evidence offered by Cleophas Kamitatu in *La Grande Mystification du Congo-Kinshasa*: "Can one deny corruption when Colonel Njufula, Chief of Staff of the Armed Forces owns four apartment buildings? . . . . And what about Colonel Biumba, in charge of Mobutu's personal guard: how many buildings does he own? What about his *train de vie?* What foreign politicians, even holding the functions of President of the Republic, could live as comfortably? Does the public know that, following Mulele's execution, each participating officer received 3,000 zaires from the President?" [32] The military, on the whole, acts as an economic entrepreneur. Like their civilian compeers the senior officers constitute a class of *parvenus* whose life-styles and standards of living mark them off sharply from the masses. Their material stakes in the status quo insure their commitment to the regime.

To stress the saliency of "class" as a reference group for both civilian and military elites is not meant to minimize their susceptibility to other kinds of identifications, ethnic, regional, or generational. Our argument, rather, is that for the time being, class affiliations remain the single most important source of cohesion between them. Nor does this situation necessarily imply a lack of competitiveness. Stability depends on the continuing ability of the system

32. C. Kamitatu, *La Grande Mystification du Congo-Kinshasa* (Paris, 1970), p. 227.

to meet the demands of its supporters, civilian and military, for payoffs, kickbacks, and personal favors. In this respect the attitudes prevailing within the army are strikingly reminiscent of civilian attitudes: "Each man comes first, then his unit, then a particular leader with special pull with Mobutu." [33] One finds in the officer corps the same attitude of self-centeredness, cynicism, and short-run calculation of benefits that Comeliau has identified in the upper ranks of the civil service, a situation which, as he recognizes, raises major obstacles in the way of economic and social breakthroughs.[34] Given the comparatively low standards of professionalism of the ANZ, patronage and "prebends" are, for the time being at least, the most effective ways of keeping the military loyal to the regime. Whether the campaign of "authenticity" currently under way will suffice to still the simmering discontent of the new generation of noncoms is as yet uncertain.

Burundi: The ANB as Ruling Caste

Although the conditions that precipitated the advent of army rule in Burundi are strikingly similar to those prevailing in Zaire—an extremely low level of institutionalization, an anemic party apparatus, and a recurrence of ethnic violence, the case of Burundi introduces a major variant in the pattern of praetorianism we have just discussed. Being much closer to the ruler type discussed by Perlmutter,[35] the ANB is more directly vulnerable to political tensions and ethnic rivalries, a situation further aggravated by the extreme saliency and increasingly rigid dualism of ethnic competition in what is now left of the political system. Ethnic polarities were from the outset much more difficult to manage than in Zaire, owing in part to the horizontal nature of ethnic cleavages, and in part because of the absence of an identifiable geographical center from which to manage ethnic conflict. As recent events tragically demonstrate, Micombero's efforts to civilianize his rule through a manufactured ideology and party organization utterly failed to dissipate ethnic tensions. The strategies of accommodation employed by Mobutu to solidify the loyalty of the army proved unworkable in the case of Burundi. The sheer saliency of ethnic feelings made it impossible to insulate the army from civilian politics; the paucity of economic and financial resources imposed severe limitations on the scope and efficiency of political patronage; and the relatively small size of the armed forces made it impractical to introduce functional divisions within the ANB as a means of balancing ethnic rivalries.

33. From a source who prefers anonymity, in a personal communication, 22 February 1972.
34. C. Comeliau, *Conditions de la Planification de Development: L'Exemple du Congo* (Paris, 1969), p. 216.
35. Perlmutter, "The Praetorian State and the Praetorian Army," *op. cit.*

The conditions of massive and systematic ethnic violence recently experienced by Burundi (resulting in an estimated one hundred thousand deaths) make it impossible at the time of this writing to identify the emerging contours of the polity.[36] The orgy of counterviolence triggered by the abortive, Hutu-led uprising of 29 April 1972, has resulted in what can perhaps best be described as a situation of controlled anarchy. Ethnic identity controls the direction of violence but imposes no limitations on its perpetrators. The boundaries between army, government, and society are almost indistinguishable. Everything seems to merge into the mold of ethnicity. *Jeunesses* and noncoms, both Tutsi, cooperate in the killings in the countryside and the capital city, while students and soldiers work hand-in-hand to round up and incarcerate all literate Hutu; ministers and senior officers supervise the logistics of the operation; Micombero, meanwhile, issues statements to the press.

However extreme in their manifestations, the convulsions experienced by Burundi are by no means unique; nor is the fact of ethnic rivalry the only dimension relevant to an understanding of civilian-military relations over the last decade. Indeed, better than any of the three states under consideration, the case of Burundi vividly illustrates the mutability and variable saliency of reference-group identifications among civilian and military elites. Although ethnicity has undoubtedly been the primary axis of reference around which civilian-military relations revolved since the army first intervened, in 1965, until April 1972 ethnic identifications were periodically offset by the incidence of clan, class, regional, and generational cleavages within each of the major ethnic groups (Hutu and Tutsi). Only with the massive upsurge of ethnic violence, on a scale never before experienced, did these secondary identifications evaporate, leading to a reassertion of ethnic solidarities.

The typology in table 4 indicates in schematic terms the changes that have taken place in reference-group identifications within the army, and how these changes have in turn affected the profile of the governing coalitions.

1. Until 1965 identifications within the army primarily centered upon monarchical symbols, reflecting the residual "aura of legitimacy" gained by the Bezi branch of the monarchy as a result of its nationalist stand during the terminal stages of the trusteeship.[37]

2. As Hutu-Tutsi rivalries began to penetrate ever more deeply into the fabric of society, reducing the party, the bureaucracy, and finally parliament to a state of near paralysis, the crown emerged as the main institutional pivot around which the political life of the kingdom revolved. The refusal of the palace to acknowledge the electoral victory of the Hutu faction, in

36. See Marvin Howe's assessment of the massacre in the *New York Times,* 11 June 1972.
37. See Lemarchand, *Rwanda and Burundi,* pp. 289–324.

Table 4: *Typology of Military Interventions: Burundi, 1965–72*

Date	Origins	Actors' Reference Group	Target	Character	Outcome
Oct. 18, 1965	Hutu army and gendarmerie officers	Ethnic and "class"	King and Monarchy	Abortive coup	Limited, Indirect Military rule, followed by purge of 34 Hutu officers and soldiers, 9 gendarmes and 86 civilians
July 8, 1966	Predominantly Tutsi army officers and politicians	Ethnic and Generational	King and Government	Dynastic-Governmental coup	Dual civilian-military rule, followed by advent of new king and new government
Nov. 28, 1966	Predominantly Tutsi army officers	Ethnic and Generational	King and Monarchy	Revolutionary coup	Direct Military rule, quasi-civilianized; proclamation of Republic
Sept. 16, 1969	Hutu officers and civilians	Ethnic and "class"	Government	Rumored coup	Purge of 19 Hutu army officers and 4 civilians
Oct. 20, 1971	Tutsi-Banyaruguru officers and civilians	Regional-Clanic	Government	Rumored coup	Temporary arrest of Banyaruguru and Ganwa elements, followed by establishment of the Conseil Suprême de la République (CSR)

1965, led to a Hutu-led *abortive coup* on 18 October 1965, and to a situation where the "loyalist" faction (primarily Tutsi) of the military exercised considerable influence behind the scenes.

3. The *dynastic-governmental coup* of 8 July 1966, instigated by a temporary coalition of Tutsi army officers and politicians, brought a new king to the throne (Ntare) and a new government to office, but left the constitutional framework untouched. The result was a dual civilian-military rule in which ethnic and generational affinities provided the essential links between the king (at the time in his late teens), army men, and bureaucrats.

4. The *revolutionary coup* of 28 November 1966 was in fact a rehearsal of the 1965 coup, being directed against the crown, except that (1) its instigators were predominantly Tutsi in origin, and (2) it succeeded where the other failed. What semblance of power the monarchy still had was thus abruptly terminated with the proclamation of the Republic in November 1966.

5. The *"rumored" coup* of 16 September 1969 further accentuated the ethnic profile of the governing coalition, being followed by a major purge of Hutu officers and politicians. By December 1969 there appeared to be only two Hutu officers left, and apart from a few successful Hutu ministers, essentially designed to act as "tokens," Hutu elements were systematically eliminated from the government and bureaucracy.

6. The *"rumored" coup* of 20 December 1971 shifted the bases of conflict away from a straight Hutu-Tutsi confrontation to an intra-Tutsi conflict, with regional and clan affinities representing the two most salient reference-group identification for the northern Banyaruguru and the southern Bururi-centered Hima factions.

Table 5: *Ethnic, Regional, and Clan Origins of CSR Members:*
Burundi, 1971

| | Ethnic Origins | | | Regional-Clanic Origins | |
| | | | | Hima | Banyaruguru |
Rank	Tutsi	Hutu	Ganwa	(South)	(North)
Lt. Colonel	1				1
Major	15	2	2	6	9
Captain	6			*	*
Lieutenant	1			*	

(*) Unknown.

The data in table 5 indicates the distribution of ethnic, regional, and clanic affiliations within the *Conseil Suprême de la République* (CSR), a junta-type organization set up by Micombero on 20 October 1971, and composed entirely of army officers. Officially the role of the CSR was "to counter all tendencies likely to endanger national unity and peace . . . to give its opinion on the selection, maintenance in office or replacement of the persons responsible for the stewardship of public affairs and to insure discipline in all State organs." [38] In setting up the CSR, Micombero optimistically assumed that the sense of corporate identity of the officer corps would produce consensus and cohesiveness where none previously existed. The dismissal of his predominantly civilian government on 29 April, coinciding with the Hutu uprising, gave the army and the *Jeunesses Révolutionaires* a free hand to repress the insurgents. Meanwhile the massacre of 450 Hutu troops, in early May, on the orders of the ANB high command, has since resulted in an all-Tutsi army. Once again ethnicity reasserted itself with a vengeance.

More than ever the role of the military in Burundi is essentially a prophylactic one. Micombero's stated objective is to rejuvenate and purify the nation ("mettre à la retraite les essouflés, et punir sévèrement les nocifs"), but in practice purity is achieved through *épuration*. The scale on which this surgery has been performed has deprived the country of its ablest sons. In the name of the slogans associated with the concept of "national revolution" what little potential there first existed to operate an economic and social revolution has been frittered away. Praetorianism in Burundi has brought a new caste to power, yet its contribution to nation-building (unlike that of its historic predecessor) scarcely goes beyond the preservation of its own economic and social privileges.

Rwanda: The GNR as Citizen Army

The case of Rwanda deviates most obviously from the varieties of praetorianism exemplified by Zaire and Burundi. Although civilian and military elites are, as in Burundi, drawn from the same ethnic stratum, the army remains under civilian control. Moreover, Hutu rule in Rwanda means the rule of politicians claiming to represent ninety-five percent of the population; army rule in Burundi is not only equated with Tutsi rule, but, for the Hutu masses, with minority rule of the worst kind.

Owing to the circumstances of its birth, the GNR was from the very outset animated by sentiments which sustained and reinforced the revolutionary aspirations of the civilian elites. Its ethos, in a way, is that of the nation-in-arms. Defending the state against its domestic and foreign enemies is part of the more fundamental reconstruction of society envisaged by the civilians. The army's commitment to the principle of civilian supremacy is more than just a reflection of a basic concurrence of ethnic interests between military and civilian elites. It is the logical consequence of its adherence to revolutionary ethics. Since the revolution of 1959 was aimed at the obliteration

38. *Flash-Infor*, Bujumbura, 22 October 1971.

of minority (i.e., Tutsi) rule, army rule would merely perpetuate in a new guise the worst feature of *l'ancien régime*.

Once this is said, attention ought to be drawn to the somewhat ambiguous position of the GNR in the political system. For one thing, the coup that brought the Hutu elites to power (the so-called Gitarama coup of January 1961), though essentially bloodless and carried out exclusively by Hutu civilians acting in accord with the Belgian authorities, may prompt the army to resort to a similar procedure; hence the deliberate efforts made by President Kayibanda to stress the position that the Gitarama coup was not really a coup but a revolution. Any attempt by the military to engage in coup-making is viewed as contrary to both the spirit and the methods of the revolution.

The disposition of the GNR to intervene also stems from the social and cultural characteristics of the armed forces. Among the officers several continue to enjoy the privileges and social status traditionally associated with the old *bakonde* families of the north, being themselves of *bakonde* origins. Their social and regional origins, coupled with their material stakes in preserving a system of land ownership that some might characterize as "feudal," imparts a somewhat "reactionary" coloration to their attitudes. The vested interests of the officers in maintaining the traditional (i.e., pre-Tutsi) social and economic organization of the northern region is certainly one major reason why Kayibanda has consistently avoided all initiatives which would imply a redistribution of wealth and social status in the north.

Reference-group identifications within the officer corps are not only economic and social but regional. Like the bulk of the troops, they are predominantly recruited from the northern prefectures, partly on the assumption that the war-like northerners would make better soldiers, and also because of the vertical absence of alternative channels of mobility in the northern regions. Since the majority of the civilian elites originate from the central (i.e., Gitarama) and southern regions, many of the cultural discontinuities which separate north from south are replicated in the differences of attitude discernible between civilians and military elites. The more salient the incidence of the north-south cleavage on national politics, the greater the chances of military intervention. Thus if the characteristics of the praetorian model are clearly not applicable to Rwanda, the seeds of praetorianism are nonetheless present in the political system.

The strongest inducements for military intervention are likely to originate from the newly perceived threats posed to the security of Rwanda by Burundi's military ethnic posture. With each society moving steadily in the direction of ethnic homogeneity at the elite level, chances of external confrontations will increase correspondingly, with the Zairian army perhaps adding yet another dimension to its arbitrating functions. Let us, then, shift our ground for a moment to look at the external matrix of civilian-military interactions.

The External Matrix of Interaction

So far we have attempted to locate our conception of military behavior within the framework of a contextual analysis of civilian-military relations emphasizing the *domestic* forces that prompt the military to act as a "class," ethnic, or regional elite. Let us now recast the argument in a somewhat wider perspective, focusing upon the *transnational* forces at work in each state.

Reference-group identifications within the military, though evidently conditioned by threats emanating from the domestic arena, must also be viewed in the perspective of the "external" forces at work in each state—namely (a) the spillover of antiregime forces into adjacent national arenas, and (b) the conversion of what might be referred to as "extraregime" forces (i.e., essentially mercenaries) into antiregime forces.

The Spillover of Antiregime Forces.——In each state political violence has generated massive involuntary migrations of individuals and groups into adjacent territories. Each state has thus become host to a refugee community from one or the other of its neighbors: from 1959 onwards Tutsi refugees from Rwanda have sought asylum in Burundi and Zaire; during and after the Congo rebellion (1964–65) a sizable minority of refugees from the Kivu and Katanga provinces has found refuge in Burundi; more recently tens of thousands of Hutus have left Burundi to seek protection in Rwanda and Zaire.

All are in a sense "political" refugees, inasmuch as their exodus stems not from natural but political calamities. As is often the case in Africa, the political aspect of the problem cannot be easily separated from its ethnic or cultural dimensions. Yet there are notable differences in the extent to which the *ethnic identity* of refugee groups may be said to have "caused" their exodus: refugees from Rwanda were primarily of Tutsi origins, and those from Burundi of Hutu origins. Refugees from Zaire, however, were of mixed ethnic origins. Another variable to consider has to do with the *timing* of their exodus. Chronologically, the flight of Tutsi refugees from Rwanda preceded the flight of refugees from Zaire into Burundi, and both preceded the exodus of Hutus from Burundi. The element of dyssynchronization between the political crises that prompted these involuntary migrations, and the crises that caused the rise of antiregime forces in the host community is crucial to an understanding of the linkages established between refugee organizations and antiregime forces in each state. So, also, is the ethnic factor. Just as the differential timing of political crises in each state provided the breathing space necessary for the working out of tactical alliances between refugee groups and domestic opposition forces, the degree of ethnic affinity between them has been of crucial importance in determining the strength and durability of their alliances.

The effect of these variables on the behavior of military elites is nowhere more evident than in the case of Burundi. The cultural or ethnic characteristics of the refugee communities settled in Burundi have had a profound impact on the threat perceptions of military elites. Conversely the perceptions of the refugees with regard to the Burundi elites, both military and civilian, have also played a part in shaping their responses.

For the sake of analysis one might conveniently distinguish three main phases in the patterns of interaction that have taken place over time between refugee communities and national actors:

1.   From 1962 to 1966 the closest area of interaction was between the Tutsi refugee community from Rwanda and the Tutsi community of Burundi, each tending to rely on the other's assistance to attain its own political ends.[39] While the Tutsi refugees courted the support of their Burundi kinsmen to fight their way back into Rwanda, the latter used the Rwanda refugees to gain leverage against the Hutu elites. While adding significantly to the emotional quotient of the Hutu-Tutsi problem of Burundi, the growing ties of interdependence between the two communities enabled the refugees to set up a paramilitary organization (the so-called *Armée Populaire de Libération Rwandaise*) for the specific purpose of "liberating" the homeland. Besides increasing the chances of a global Hutu-Tutsi confrontation—a confrontation which the security forces of Burundi, given their limited response capabilities at the time, could not effectively arbitrate without causing ethnic revolts within the army—the setting up of a rival refugee military organization was bound to generate feelings of apprehension among army officers. The threats which this situation posed to the integrity of the Burundi army was undoubtedly instrumental in Micombero's seizure of power in July 1966. The situation was further aggravated by the arrival in Burundi, in December 1965, of some five hundred *inyenzi* hard-core fighters. Most of them came from the Congo where they fought at the side of the Congolese rebels before being pushed back by the counteroffensives of the ANC. Thus by 1966, the chief concern of the Burundi military elites was to avoid at all cost a civil war which, given its ethnic dimensions, could also spell the disintegration of the armed forces.

2.   From 1966 to 1971 the boundaries of conflict shifted from Hutu-Tutsi rivalries to intra-Tutsi rivalries, with the Rwanda refugees, for cultural and historical reasons, throwing their weight behind the Tutsi-Banyaruguru faction against the Tutsi-Hima (also known as the Banyabururi) faction. Given the preponderance of Hima (or Bururi) elements within the army, however, the power struggle rapidly transformed itself into a civilian-military struggle within the Tutsi stratum. The events of late 1971, leading to the temporary incarceration by the army of a number of Banyaruguru elements holding

39.   For further information, see Lemarchand, *Rwanda and Burundi*.

key cabinet positions, were the logical consequence of the perceived threats which the alliance of the Rwanda refugees with the Banyaruguru faction posed to the clan and regional interests of the officer corps, by then mostly composed of Bururi elements.

3.   From late 1971 to April 1972, there occurred a partial reversal of alliances, eventuating in an increasingly close working relationship between the Congo refugees and the Hutu population of Burundi. Until approximately 1966, the Congo refugees were generally willing to follow the tactical alliances of the Rwanda faction, the assumption being that the Tutsi elites of Burundi were their safest bet to set up a "privileged sanctuary" into Burundi from which to launch guerrilla operations into the Congo. The obduracy of the Burundi military, however, caused the Congo refugees (many of them former *Simbas* driven into Burundi by the ANZ) to switch sides suddenly, and to capitalize upon the latent discontent of the Hutu population to unleash a massive attack against all Tutsi elements. The result was the military-sponsored anti-Hutu repression of May and June 1972, amounting to a major hecatomb of Hutu elements, in turn prompting one of the largest exoduses of refugees thus far experienced by any of the three states.

A more sustained exploration of the impact of antiregime forces on military behavior is not possible here. What evidence has been offered, however, is enough to disabuse us of the notion that the attitude of the military is solely conditioned by internal variables. As much as the ethnic map of each national society, the ethnic and cultural characteristics of antiregime forces, the accommodation of such forces to the political environment of the host society, and the capabilities of military actors to cope with the threats emanating from exogenous forces, were all instrumental in shaping the responses of the military. A somewhat similar conclusion emerges from an examination of the impact of "extraregime" forces on military attitudes.

The Impact of Extraregime Forces.——"Extraregime" forces include a variety of exogenous pressures and instrumentalities. These may range from diplomatic pressures to covert activities of the kind associated with the CIA to the hiring of foreign mercenaries. As Ruth First justly emphasized,[40] few other states have been more directly exposed to such influences than Zaire; in few other states, in any event, has the military been more heavily and cumulatively dependent upon the "suspensive" veto of Western embassies, the financial backing of CIA agents, and the military and logistical support of mercenary forces. The part played by CIA funds in enabling Mobutu to pick up the pieces of the ANC and the effect of diplomatic pressures on Mobutu's decision to seize power in 1965 have yet to be written. Even though

40. First, *Power in Africa,* pp. 411ff.

the full story may never be disclosed, there can be little doubt that these external pressures played a significant part in enhancing the response capabilities of the ANC and hence its aggregate "weight" in the political system. By the same token, one might also argue that the European encadrement of the GNR at the time of the so-called Bugesera invasion played a major part in the successful repulsion of *inyenzi* commandos. And the moderating influence of Belgian military advisers in both Rwanda and Burundi must have been a factor of no small significance in averting major border clashes in the mid-sixties. In the absence of reliable information, however, these hypotheses are little more than educated guesses.

Yet enough is known of the recent history of white mercenaries in Zaire to suggest some tentative contrasts and similarities with the transnational forces already described. One of the key lessons to be drawn from the involvement of white mercenaries in Zairian politics is that, to quote Machiavelli, "if anyone supports his state by the arms of mercenaries, he will never stand firm or sure as they are disunited, ambitious, without discipline, faithless." By stressing the attitudinal orientations of mercenaries, Machiavelli's warning helps us understand a major characteristic shared by mercenaries with refugee groups, namely, their tendency to re-orient their psychological identifications in the light of the political changes taking place in the host society. As long as the mercenaries defined their role in the Congo in strictly military terms, that is, in occupational terms, their role has essentially been a supportive one vis-à-vis the civilian-military elites of the Congo. Indeed the expectation of continued military support from white mercenaries was probably instrumental in Mobutu's decision to seize power in 1965. What Mobutu did not expect was the drastic change in the attitude of the mercenaries after Tshombe's fall, in 1965, and even more markedly, after his arrest and surrender to the Algerian authorities in 1967. Nor, of course, did he expect the former Katanga gendarmes incorporated into the ANC to join hands with white mercenaries to engineer a forcible seizure of power. This situation suggests a curious re-enactment of the process of re-identification analyzed in our discussion of antiregime forces. Indeed, by 1967, the mercenaries had converted themselves from an extraregime into an antiregime force.

The shift of identifications produced by the changing fortunes of the Tshombist clique can be observed at two levels: (1) at the level of cultural differences within the mercenary forces, taking into account not only the national origins of the mercenaries (French, Belgian, South African, etc.) but, even more importantly, the "settler" and "non-settler" groups; and (2) at the level of cultural or regional differences within the Zairian army. That the strongest impulse to challenge Mobutu's rule happened to originate from "settler" mercenaries (like Schramme) and former Katanga gendarmes, is not coincidental. As much as their residual sense of loyalty to Tshombe, the long period of European encadrement experienced by the Katanga Army was a powerful inducement for the ex-gendarmes to join hands with the mercenaries.[41]

The immediate effect of this conversion process on the ANC was to significantly weaken its overall military capabilities. Ultimately, however, the defeat of the mercenaries undoubtedly strengthened its morale. As Willame points out, the story of the Katanga units "tragically ended with the Kisangani mutiny of 1967, after which the bulk of the Katangese gendarmes were purely and simply massacred by ANC troops"; on the other hand, "their adventure produced the reverse effect of the one they had expected: The ANC did not surrender. After the departure of the defeated mercenaries, the self-confidence of the Army, as well as its arrogance, was completely reestablished." [42]

Though they frequently experienced the weight of "extraregime" forces, Rwanda and Burundi were both fortunate in being spared the involvement of foreign mercenaries in their domestic politics. It is interesting to note, however, that among the several motives that prompted the Burundi army to intervene in 1965 was the persistent, and probably not unfounded, rumor of Mwambutsa's decision to hire West German mercenaries. And one of the motives invoked for the incarceration of ex-king Ntare, in 1971, was his alleged intention of introducing West German mercenaries into Burundi. Whether there is any truth to these allegations is beside the point. What is significant is that in each case perceived threats to the security of the army were instrumental in prompting its intervention.

In view of what has been said up to this point, there seems no reason to assume that changing identifications within the military are to be found only among African armies. The evidence from the mercenary forces hired by Mobutu suggests a phenomenon similar to what can be observed among the indigenous military elites of each state. If further evidence were needed, one might find it in the changing allegiances of the French army after World War II. In Africa as elsewhere, in short, the posture of military elites reflects, in varying degrees, the contextual realities of their sociopolitical, or ethnic, environment.

Conclusion

In none of the three states under consideration is there any evidence that the military is either willing or able to engage in the manifold tasks associated with nation-building. With the qualified exception of Rwanda, the aim at best is "state-building" rather than "nation-building," not the productivity

41. For additional information, see the excellent discussion of the 1967 revolt in *Congo 1967* (Brussels, 1968), pp. 365–420.
42. Willame, *Patrimonialism and Political Change in the Congo,* chap. 3, p. 37.

of political resources through social mobilization but the monopoly of the means of coercion.

Part of the explanation for this state of affairs lies in the character of the Belgian colonial legacy and the effects of this legacy on the political evolution of each state. Nowhere else in Africa has a colonizing power shown greater indifference to the development of professional skills within the military then in these three states; nowhere else has the repressive capacity of a colonial army been given higher priority than in the Congo. The extremely low level of institutionalization of civilian structures—reflecting an even more fundamental feature of the Belgian legacy—further enhanced "security" as the prime function of the armed forces in the period immediately following independence. Moreover, the "spillover" of internal rebellions across their national boundaries, whether in the form of refugees seeking asylum or "liberation forces in exile," has often led to active resistance on the part of these exogenous elements to the policies of the host country, and in some instances (as in Zaire and Burundi) to the direct intensification of domestic rivalries. The result has been to create an atmosphere of obsidianal fear within the host community which also has served as justification for further emphasizing the security aspects of military rule.

Another explanation is found in the characteristics which some have attributed to praetorianism, regardless of the environment. Because of their congenital aversion to politics and politicians, it is sometimes asserted that once in power army men seldom bother to create the conditions required for self-sustained development. If the evidence from Zaire and Burundi is at all conclusive, however, the problem is perhaps not so much that officers shirk from assuming the role of political organizers, but that their style of political entrepreneurship is often self-defeating. Rhetorical allusions to "authenticity" can scarcely conceal the hollowness of the Zairian polity. Micombero's style is authoritarian and paternalistic: loyalty to the establishment is rewarded, and dissent severely sanctioned; the scope of initiative allowed to the bureaucracy and cabinet members has constricted to the point where no major decision can be made without prior consultation with Mobutu; and in case the prospects of material benefits do not suffice to induce compliance, force most certainly will. In his attempt to escape from the politics of ethnicity of the post-independence period Mobutu has in a way turned the clock back to pre-independence politics, to a situation where there are no politics.

Perhaps the most plausible answer is that the military is no more willing or able to identify with the interests of society as a whole than the civilians. The position of the military, after all, is not nearly as insulated as some would have it. Armies are but one among several competing interest groups. Self-identifications within the military reveal the same kind of mimetic changes that others have observed among civilian groups, except that the range of possible self-identifications is made broader still by the institutional divisions characteristic of military establishments. In their pursuit of power and privilege armies are not only competing with civilians, but different sections of the army are usually competing among themselves. In these conditions the assumption that professionalization will automatically bring about a developmental millennium seems scarcely tenable. Nor can one argue that the ethnic factor is the only relevant source of change both within and outside the military. We have seen how ethnic imbalances within the officer corps may precipitate armed revolts within the army (as in Burundi); but we have also drawn attention to a variety of other contextual variables, regional, clanic, and generational, each cutting across ethnic cleavages. Thus any attempt to read into African military behavior an ethnic version of the Latin American middle-class coup[43] seems equally misguided.

Reformulating Von Clausewitz's remark that war has its own grammar but not its own logic, one might say that African military behavior has both a grammar and a logic of its own. Yet it is a logic which postulates a far greater number of contextual referents than most observers have been willing to concede, and hence also a far greater number of possible permutations. Our main concern in this article has been to clarify the postulates of military behavior—to show how different threat perceptions may activate different solidarities. There remains the more arduous task of systematically relating these postulates to specific policy outputs and system changes.

43. José Nun, "A Latin American Phenomenon: The Middle Class Military Coup," in *Trends in Social Science Research in Latin American Studies* (Berkeley: 1965), pp. 55–99.

PART IV

Asia

GETTING OUT OF POWER:
THE CASE OF THE PAKISTANI MILITARY

Byron T. Mook

I

Political scientists have produced a vast literature on military coups. One is struck, after having examined this material, by the disproportionate emphasis the authors have placed on a single set of questions. Why has the military taken power in the first place? What social and/or economic factors have caused an ongoing political system to break down? What roles have military history and/or interservice rivalries played in precipitating military intervention?

Very seldom, however, does one come across examination of an equally important question: how does a military government get out of power, or how does it try to do so? The reason for the relative dearth of literature on this subject is clearly the scarcity of successful cases. Peaceful and durable transitions from military to civilian rule have been few. The purpose of the analysis which follows is to describe some of the constraints under which military governments operate in trying to get back to the barracks. The case to be looked at in detail is Pakistan.

No one who followed the events of 1971 on the Indian subcontinent can have remained completely uninvolved in them. Massive repression, ten million refugees, invasion, and war provoked heated denunciations and counter-denunciations by the Pakistani and Indian governments. Outstanding issues between those governments remain unsettled even today. General lessons are obviously difficult to abstract so soon after such a cataclysm. Still, the long period of military rule in Pakistan provides an almost ideal study of how difficult it is for the armed services of a country to turn effective political power back to civilians. Military concessions to demands for increased popular participation provoked army reaction, which in turn provoked more radical political demands, which in turn provoked more army reaction, which finally culminated in a (successful) political attempt to break up the state.

It is easy to argue all this with the benefit of hindsight, of course. Nevertheless, a strong case can be made that there really was a point of no return, probably about 1966, after which it would have been almost impossible for the military to get out of power without an upheaval of the type which occurred. The events of 1971 were in many ways the inevitable denouement to a strategy which never had much hope of success.

The main argument of the paper is that thirteen years of military rule had created impossibly narrow political parameters for the military. When General Yahya Khan took over the military government from President (formerly General) Ayub Khan in 1969, he had very few options for removing the military from power. A peaceful transition to civilian control would have required a military leader with both a very solid base in the armed forces and a farsighted political vision. Yahya Khan had neither. All this is not to say, however, that either he or his military colleagues were not well-intentioned on the question of getting themselves back to the barracks. There is strong evidence, in fact, that they were.

II

What were the parameters within which the military government operated? An examination of this central issue requires a brief look at the most salient characteristics of modern Pakistani history.

First, the areas of British India included in the new state of Pakistan had never had strong colonial institutions. As late as 1947, the British government had assumed that it would be able to avoid partition. It had, therefore, never set up separate public structures on which Muslim League leaders could practice self-government. When last-ditch negotiations to avoid a breakup finally failed, the British simply decided that the best policy was to get out quickly. They therefore gave the two governments seventy-two days to make final arrangements for independence. Such a short lead-time created hardship even for India, but for Pakistan it was a disaster. The new government in Karachi was suddenly thrown on its own with inadequate buildings, supplies, and personnel. Only about 150 civil servants from the elite "Indian Civil Service" decided to join Pakistan.[1]

1. For discussions of partition and its immediate consequences, see V. P. Menon, *The Transfer of Power in India* (Princeton: Princeton University Press, 1957); Maulana Abul Kalam Azad, *India Wins Freedom* (Calcutta: Orient Longmans,

Second, Pakistan had never had a strong political party through which popular demands could be channeled. The Muslim League before 1947 was strongest in parts of the subcontinent which later were included in India. Mohammed Ali Jinnah, the "father" of the country, was originally from Bombay, and Liaqat Ali Khan, the first prime minister, was from the United Provinces. As a result, the men who were expected to lead the new state had no firm political bases or constituencies in it.

And third, given the weakness of the nationalist movement in precisely those areas included in the new state, Pakistan was never able to form a strong national identity. Geographical division into east and west "wings" aggravated this situation. Bengalis claimed almost from the beginning that Punjabis, Sindhis, and Pathans discriminated against them in terms of administrative jobs, political power, and development resources. The new state was not even awarded the major cultural centers around which a sense of national pride and commitment might have coalesced. Lahore did go to Pakistan, but all the secondary cities of the Punjab, and, most important, Calcutta, went to India. This geographic allocation was a symbolic disaster to leaders of the new country, who had counted on getting a Pakistan which included all of the Punjab and all of Bengal, at a minimum. The only source of national unity in the early years was hatred of a foreign enemy—India.

Without any of the major prerequisites for either statehood or nationhood, therefore, Pakistan set out on an independent path. But almost immediately, the new state was subjected to a series of severe political shocks. First was the trauma of a population exchange unprecedented in human history. Estimates are that approximately seventeen million people left their homes and tried to cross the new international borders. Of these, several millions were murdered, and those that succeeded in reaching their destination had to be resettled. Second, Pakistan lost a war with India over the disputed state of Kashmir. And third, the ultimate blow was the death one year after independence of Mohammed Ali Jinnah. Three years after that, Liaqat Ali Khan was assassinated.

The results of this unpromising beginning and multiple catastrophes were predictable. Without leaders, parties, or consensus on the rules of the game, Pakistan immediately lapsed into a period of chaotic parliamentary politics. A succession of governments appeared, none of which was able to make a concerted assault on the major problems confronting society. Basic issues such as the relationship between executive and legis-

1959); Leonard Mosely, *The Last Days of the British Raj* (New York: Harcourt, Brace and World, 1961); Penderel Moon, *Divide and Quit* (Berkeley: University of California Press, 1962); Chaudhury Khaliquzzaman, *Pathway to Pakistan* (Lahore: Orient Longmans, 1961); and Wayne A. Wilcox, "The Economic Consequences of Partition: India and Pakistan," *The Journal of International Affairs* 18, no. 2.

lative branches, and between the center and the provinces, remained unresolved. Political leaders could not even agree on a constitution until 1956.
The main beneficiary of such a confused situation was, of course, the civil
service. In the absence of strong political direction or control, it ruled. As
we look back on the first years of Pakistan, it is clear that the only
reason the state survived at all was because of a very small group of very
able men at the top levels of administration.[2]

The military, meanwhile, was unhappy with a government which could
not provide it with what it felt was the material necessary to do its job. It
had lost its first confrontation with India in 1948. It was short on men,
very short on supplies, and frustrated by its inability to take Kashmir. This
malaise surfaced for the first time in 1951 in the form of an attempted
military coup. The government moved to moderate this discontent by sign-
ing a national defense and arms supply agreement with the United States in
1954.[3] But four years after that, ostensibly in order to put a stop to civilian
chaos in government, the military took over anyway. It is probable that
General Ayub Khan, the commander-in-chief, also feared an electoral victory in 1959 by H. S. Suhrawardy, a Bengali politician, who might have
created the first stable civilian government Pakistan had ever had. Ayub
was probably not opposed to such an outcome per se, but Suhrawady was
a Bengali. Since most of the military was composed of non-Bengalis, Ayub
feared that military demands would be as unheeded in the future as they
had been in the past. At the same time, a further stimulus to action was
a spate of rumors of a possible military coup by his own junior officers unless
he took strong action.[4]

The military came to power, then, largely as a response to what it saw
as the weakness and incapacity of civilian government. Ayub promised
that he would turn power back to civilians only after new institutions had
been perfected and new leadership recruited. His first official act, however,
was to ban all political activity. He outlawed political parties and disqualified
selected political leaders from holding public office. The result of such
policies was to continue the dominance of the civil service, of course,
and to alienate precisely those civilians who might have provided alternative
national leadership.

2. The best analyses of this period are in Khalid Bin Sayeed, *Pakistan: The Formative Phase* (Karachi: Pakistan Publishing House, 1960); Keith Callard, *Pakistan: A Political Study* (London: Allen & Unwin, 1957); and G. W. Choudhury, *Constitutional Development in Pakistan* (Lahore: Orient Longmans,
   1959).
3. For a critical view of the role of the United States in concluding this agreement, see Selig Harrison, "Case History of a Mistake," "Cost of a Mistake,"
   and "Undoing a Mistake," *The New Republic,* 10 August 1959, 24 August 1959,
   and 7 September 1959.
4. The coup is discussed by Wayne A. Wilcox in "The Pakistan Coup d'Etat of
   1958," *Pacific Affairs,* Summer 1965.

Ayub apparently saw his first job as that of economic development. To that end, he opted for a set of policies based on what some have called "the social utility of greed." He minimized the role of the state in Pakistani economic affairs. The growth of a progressive middle class, a bourgeoisie which would be the foundation on which future democracy in Pakistan could build, was one of his major goals. Even though many of his military colleagues were landed aristocrats, he himself clearly did not believe that feudal relationships of the kind then widespread in Pakistan were compatible with stable democratic civilian government.

His system of "basic democracies," introduced in 1962, further encouraged the growth of this new middle class. Indirect elections to new local, provincial, and national assemblies were expected to stimulate new political leadership. Ayub himself entered politics under the new system. He became the president of Pakistan rather than its commander-in-chief. He resurrected the Muslim League and became its president.

The prime beneficiaries of these policies are easy to identify. *First,* as noted, was the civil service, which continued its dominant role in Pakistani politics. The new representative system centered around "basic democracies" was not yet strong enough to act either as a mechanism of policy direction or control over the administration. *Second,* business thrived under a government which encouraged its growth. It became extremely concentrated. In 1968, the chief economist of the Planning Commission disclosed that sixty-six percent of the entire industrial capital of the country was concentrated in the hands of twenty families, who also controlled eighty percent of the banking and ninety-seven percent of the insurance.[5] The national economic growth rate was one of the highest in the world. Links between business and the military government were forged by the many former military officers who took over top posts in large companies and state corporations upon retirement. A *third* group to benefit was the rural elite. Ayub allowed them limited political participation in politics through the "basic democracies," and they profited from general economic growth and the "green revolution" in agriculture. Rural power relationships were left mostly undisturbed. There was very little land reform. And finally, the *fourth* group to be pleased with the system was the military. It was getting the resources it felt it needed, usually about sixty percent of the national budget. After the 1965 war with India, which Pakistan also lost, this guarantee of continued military budgets became very important. Had the 1965 war resulted in a Pakistani victory, an argument might be made that the military would have been less committed to continued rule and might have been willing to return to the barracks more gracefully.

And whom did Ayub's policies leave out? First, of course, all of the

5. Cited by Wayne A. Wilcox in "Pakistan: A Decade of Ayub," *Asian Survey,* February 1969, p. 91.

prominent politicians from the pre-coup period were excluded by law. And second, exactly those social classes which were becoming mobilized by economic and educational expansion were shut off from meaningful political participation. The exercise of power under the military government was very exclusive and very informal. The military, the civil service, business, and selected rural elites ran the country.[6]

III

Having said all this, however, one should stress again that Ayub's purposes were laudable. What he wanted to do was to buy time—i.e., to keep the lid on political demands while the socioeconomic foundations for democracy, as he saw them, were being laid. Whether he could have succeeded without facing the problem of "regionalism" in the country is a moot point. The fatal weakness of his strategy was that he did not preempt or attempt to blunt this highly emotive and symbolic issue. That single failure eventually destroyed both him and Pakistan.

Some more detail about this argument is necessary. The "regional" issue had bedeviled the country ever since its formation. In 1952, for example, long before Ayub came to power, students in Dacca had been killed by police during riots over government support for the Urdu language. In 1954, there had been a ballot-box uprising in the East, an electoral victory for Bengali nationalists, which had demolished the old Muslim League and led to central government intervention in East Pakistan. Even in 1961 and 1962, the military government had to close down universities in the East because of widespread student protest over the continued detention of H. S. Suhrawardy The solution for Ayub was clearly not to do what he did. His government continued to spend more than half of the national budget on defense, thereby benefiting a military establishment which was approximately ninety percent from the western "wing." He did not significantly increase the number of top Bengali civil servants in the nation, only about fifteen percent of the total. In addition, he made some symbolic policy decisions which only inflamed Bengali sensibilities. He temporarily took Bengali songs off Radio Pakistan, for example, because he believed he was thereby facilitating the acceptance by the people of Bengali of the idea of Pakistan. Whenever Bengalis protested, Ayub tightened his grip still further.[7]

6. A good discussion of this period is contained in Karl Von Vorys, *Political Development in Pakistan* (Princeton: Princeton University Press, 1965). On "basic democracies" particularly, see in S. M. Z. Rizvi, ed., *A Reader in Basic Democracies* (Peshawar: West Pakistan Academy for Village Development, 1961).
7. The best analyses of the "east-west" problem in Pakistan are contained in Lawrence Ziring, "East Pakistan and the Central Government" (Ph.D. diss., Columbia University, 1961); Government of East Pakistan Planning Department, *Economic Disparities Between East and West Pakistan* (Dacca, 1963);

IV

Even such a brief examination of the ten years of military rule suggests the general parameters within which the military operated as it began to think seriously about turning power back to civilians. The first and most major problem, of course, was the scarcity of politicians who were both acceptable to the armed forces and also able to rally significant sections of the population behind them. This issue was particularly acute in East Pakistan. By 1968, the old politicians, the old power brokers, who had constituencies there and who had played the parliamentary game prior to the coup were gone. Ten years had intervened since the abrogation of the first constitution. H. S. Suhrawardy and Fazlul Huq, the most prominent Bengali leaders, were dead.

In the places of the old guard, all over the country, had risen a group of younger men, of a completely different political generation. These people had not been through the nationalist movement and, most important, had never operated within a parliamentary system of government. In most cases, the interests they represented were linguistically or regionally based. Ayub had not allowed party organizations, after all. Elections had been indirect. The result was that these men were really leaders of "movements" rather than of "parties." Whenever they criticized the military, they were jailed and thus became symbolic martyrs. The more they were martyred, the more popular they became, the more radical, and usually the more demagogic. In short, they were exactly the kind of politicians whom the military found unacceptable.

This stimulus-response syndrome is perhaps best illustrated in the career of Sheikh Mujibur Rahman, the leader of the Bangladesh movement. He had been in and out of jail since 1952, even before the military coup, over the regional "disparity" issue. He therefore had sufficient credentials for martyr status, and by 1966 had become one of the major spokesmen for Bengali regionalism within Pakistan. In that year, he enunciated his famous six points concerning the future of the country.[8] These included demands for a weak central government, taxation powers to be delegated to the provinces, and the right of Bengal to have its own paramilitary force. The Ayub government naturally found such demands unacceptable and quickly put the Sheikh in jail again. A year later it put him on trial for treason, then suddenly released him and summoned him as the representative of the Bengali people to West Pakistan to confer with the president. This situation,

---

Subrata Roy Choudhury, *The Genesis of Bangladesh* (Bombay: Asia Publishing House, 1972); and M. Rashiduzzaman, "The Awami League in the Political Development of Pakistan," *Asian Survey,* July 1970.

8. *Dawn,* 12 February 1966. Also see, Sheikh Mujibur Rahman, *Our Charter for Survival: Six-Point Programme* (Dacca: Pioneer Press, 1966).

in which it was necessary to recast an opposition leader in an instant from "criminal" to diplomat, illustrates Ayub's dilemma. He no longer had politicians, particularly in Bengal, to whom he wanted to turn over power.

A second constraint within which Ayub had to operate was that of concern for military unity. Whenever a military government gives some freedom to politicians, the latter begin to seek allies within the military itself. In Pakistan, Ayub was confronted with one major such politician, Zulfikar Ali Bhutto. Bhutto, a West Pakistani and the son of a prominent and wealthy landlord family, had always taken a very hawkish position on India. He had even opposed the agreement ending the 1965 war which Ayub had signed with Prime Minister Shastri at Tashkent. Both of these positions had understandably won him friends within the armed forces. He wanted power quickly, yet he was unprepared to take anything less than a fifty-fifty sharing of power with the Bengalis (a demand which Sheikh Mujib found unacceptable, since Bengalis constituted a majority of the population). So, at just the time when Ayub was trying to think about transferring power to civilian government, he also had to be concerned about his own power base. The military dilemma on this point is perhaps best illustrated by the rumor making the rounds in early 1971 that General Yahya Khan (Ayub's successor) had gone to Dacca to arrange a settlement with Sheikh Mujib, because without one he was afraid of being deposed and kidnapped by his subordinates back home.[9]

Once again, the point is that the longer a military government stays in power, the more difficult it is for it to get out. It has trouble not only in finding politicians to whom it feels comfortable in transferring power, but also in controlling its own ranks. The Pakistani military stipulated three unalterable conditions for going back to the barracks. First, the territorial integrity of the state had to be preserved. Second, there had to be at least parity in government between the two "wings" of the country. And third, any new system had to be stable enough to provide the military with enough funds to keep it healthy. The central problem in Pakistan, first for General Ayub and later for General Yahya, was that the government could not arrive at any arrangements to ensure the realization of these ends which were acceptable to the two major extra-governmental figures, Sheikh Mujib and Bhutto.

V

Why not? Many other societies marked by clear, politicized, socioeconomic cleavages have been able to work out institutional arrangements to moderate conflicts. The Netherlands, Lebanon, and Columbia are good examples. Why wasn't the Pakistani military able to devise such mechanisms

9. *Link,* 21 March 1971.

to keep the country together? An examination of this question requires a more detailed look at the specific events which led up to the 1971 crisis.

The argument here is that compromise and settlement probably were possible, but that the military leadership had certain values and commitments which made them difficult to achieve. Sheikh Mujib wanted to be prime minister of a united Pakistan. In such a position, with a Bengali-dominated National Assembly to support him, he would have been able to alter national economic and military policy in ways more favorable to East Pakistan. Development funds could have been allocated to the East in higher proportions, and more Bengalis could have been given top posts in the government and armed forces. As the leader of the majority party in the Assembly after the 1970 elections, he had a reasonable expectation that the post would be his. General Yahya himself called Sheikh Mujib the future prime minister of Pakistan.

However, there is strong evidence that Yahya did not mean what he said, and that he had no intention of turning over effective political power to Bengalis. No one yet knows exactly what Yahya and Sheikh Mujib said to each other during their fateful meetings in Dacca in March 1971. What we do know, however, is that Yahya used the period of discussion to fly in military reinforcements from West Pakistan, a fact which suggests, at least, that he was pessimistic about the outcome of the negotiations.

Why? Aside from the policy considerations noted above, a basic influence was Pathan and Punjabi views regarding Bengalis. The nationalist coalition had first fallen apart in the East, with the result that many West Pakistanis suspected Bengalis of disloyalty to the idea of Pakistan. There had been many more disorders in the East than in the West, thereby creating the impression that Bengalis were also political radicals—which some were. The army had always been weak in Bengal. From an official point of view, this last point was not an issue, since the defense of Pakistan was indivisible. But most Pathans and Punjabis regarded Bengalis as completely unsuited for military life anyway, a view which they had inherited from the British. In the British-Indian army prior to World War II, for example, approximately fifty percent of the forces were from the Punjab, twelve percent were Gurkhas from Nepal, fourteen percent were Pathans, Dogras from Kashmir, and men from the United Provinces, but almost none were Bengalis.[10] Ayub Khan himself once said of the Bengalis:

The people of Pakistan consist of a variety of races, each with its own historical background and culture. East Bengalis, who constitute

10. These figures are from Wayne A. Wilcox, *India, Pakistan and the Rise of China* (New York: Walker and Company, 1964), p. 15. Stephen P. Cohen reports that in 1956, the quota of East Pakistanis in the Pakistan Army was fixed at 2%, in "Arms and Politics in Pakistan," *India Quarterly*, October–December 1964, p. 419.

the bulk of the population, probably belong to the very original Indian races. It would be no exaggeration to say that up to the creation of Pakistan they had not known any real freedom or sovereignty. They have been in turn ruled either by caste Hindus, Moghuls, Pathans, or the British. In addition, they have been and still are under considerable cultural and linguistic influence. As such they have all the inhibitions of down-trodden races and have not found it possible psychologically to adjust to the requirements of their newborn freedom.[11]

That was the president of Pakistan speaking about the majority of his "nation." Never in all the thirteen years of military rule was there a Bengali in the inner circle of his advisers. Periodic pressures to appoint or elect a Bengali vice-president were continually rebuffed.

VI

The crisis in late 1971 was largely the outcome of this military short-sightedness. Ayub resigned from the presidency in early 1969. At that time, the military government was confronted with vocal civilian opposition. Students and workers had been demonstrating for several months in the urban centers of West Pakistan.[12] There was much talk of socialism and revolution, though Bhutto himself, who had provoked much of this agitation with his fiery speeches, was probably acting on behalf of discontented elements in the military. Had he represented only the social classes excluded from the governing coalition, of course, Ayub would have had nothing to fear. At precisely this same time, overt military dissent began to appear with the announcement of former Air Marshall Ashgar Khan that he was entering politics to contest the continued dominance of Ayub.

So Ayub retired, his "basic democracies" scheme a failure, his population growing increasingly militant, and sections of his military discontented. General Yahya Khan, his commander-in-chief, surveyed the problems confronting the country and acted just as Ayub had eleven years earlier. Yahya's first actions were to throw out the constitution, dismiss elected assemblies, and ban all political activity.

One observer of Pakistani politics described the country as "once again at the starting point." [13] On the surface, this analysis was accurate. But there was a key difference between the two situations, i.e., that Yahya had to try to transfer power back to civilians who had already been out of

11. Quoted by Col. Mohammed Ahmad in *My Chief* (Lahore: Orient Longmans, 1960), p. 87.
12. Shahid Javed Burki has analyzed this discontent in "Ayub's Fall: A Socio-Economic Explanation," *Asian Survey*, March 1972.
13. Wayne A. Wilcox, "Pakistan in 1969: Once Again at the Starting Point," *Asian Survey*, February 1970.

power for more than ten years. He was not able to enjoy a grace period during which he could experiment with new constitutional arrangements or electoral plans. His first actions as president were therefore straightforward and apparently well-intentioned. He held consultations with a variety of political leaders, seeking their views and encouraging them to speak out on the issues. He invited students to form unions in their universities. He declared that the central government would immediately embark on a program of disproportionate economic investment in the East Wing. Private sector interests which put money in the East were promised tax breaks. Perhaps most important, at least from a symbolic point of view, he announced a new series of progressive taxation measures, including significant antitrust laws, which cut at the heart of the ruling military coalition.[14]

The political results of these policies were predictable. Groups and individuals, particularly Bengalis, which had been outside the system for years, greeted the thaw with enthusiasm. A return to parliamentary government seemed imminent. The only unresolved issue was apparently the specifics of when and how the transition to civilian rule would be accomplished.

A new chill descended, however, when communal riots between Biharis and Bengalis broke out in the East. The former were Urdu-speaking refugees from India, who gradually in the 23 years since partition had acquired prominent positions in the government and economy of East Pakistan. The prospect of violence against this minority made the military wary of a return to civilian government, particularly if that government was to be dominated by Bengalis. Furthermore, at precisely this same time, Sheikh Mujib entered into negotiations with leaders of various political parties in West Pakistan. He talked especially to those men who were leaders of regional parties, who might therefore be interested in the "federal" types of constitutional arrangements which he himself favored. The military was obviously not much in favor of such decentralization either.

Yahya responded to these disturbances and political maneuverings not by repression, but by making still further concessions. In November 1969, he announced that there would be full parliamentary elections within a year, the first national elections in Pakistani history on the basis of direct one-man one-vote suffrage. He made clear that the East would have a majority of seats in the new national assembly, commensurate with its majority of the national population. He declared further that West Pakistan would be broken up into four separate provinces, long a demand of regional leaders. These were all major concessions. A month later, he went still further by announcing that three hundred top civil servants would have to "show cause" why they should not be investigated and perhaps indicted for crimes ranging from inefficiency to subversive activities.

Popular expectations continued to rise. The decision to disintegrate the

14. Ibid.

West Wing of the country to allow the formation of new provinces, coupled with Mujib's rising popularity in the East, meant that regional leaders began to appear considerably stronger than those leaders who had national constituencies or who claimed that they did. Political parties mushroomed. At the time Yahya made his announcement about the schedule for elections, there were twelve parties. By the actual time of the elections, there were twenty-five.[15]

Within three months, however, Yahya was obviously having second thoughts about the democratization process he had begun. In an attempt to prevent the possibility of national disintegration, he laid down some basic ground rules for the upcoming elections. First, he said, the new constitution which the elected assembly would draft must preserve a strong central government. Too much delegation of authority to the provinces of Pakistan would not be tolerated. And second, in order to make sure that all the new constitutional arrangements were satisfactory, he himself would have to approve them. If the new constitution were not acceptable, he threatened, it would not take effect.

Confronted with such an ultimatum, and unaccustomed to direct elections anyway, politicians throughout the country waged a vigorous and often rowdy campaign. Violence was particularly common in the East. Politicians who did not belong to the Awami League, Sheikh Mujib's party, were frequently not allowed to speak and their electoral meetings were broken up. The military does not appear to have been directly involved in this violence, though it was obvious to all observers that its sympathies lay with any party which opposed the Awami League.

When the results of the vote were counted, however, in December 1970, the Awami League had swept the East to a degree which no one had anticipated. Simply on the basis of its strength there, it had won enough seats to control the new National Assembly and to elect Sheikh Mujib as prime minister. Throughout the country there had been an apparent radicalization of political sentiment. Bhutto's People's Party had emerged as the overall winner in the West. General Yahya was therefore unexpectedly thrust into the position which many military leaders and other West Pakistanis had always feared, i.e., of having to turn over the national government to Bengalis. Even some kind of coalition between Sheikh Mujib and Bhutto, a government of "national reconciliation," the only other possibility, was not a particularly attractive one.

Yahya equivocated. On January 14, 1971, he declared that "Sheikh Mujibur Rahman is going to be the future prime minister of the country." [16] March 3 was set as the date for the opening of the new National Assembly

15. These figures are from Sharif al Mujahid, "Pakistan: First General Election," *Asian Survey*, February 1971.
16. *Dawn*, January 15, 1971.

in Dacca. But only two days before that opening, Yahya knuckled under to pressure from Bhutto and the military, and postponed the Assembly "to a later date." Bhutto insisted that he be guaranteed an equal role with Sheikh Mujib in shaping the new constitution. The military feared civilian disorders in the West if Yahya went ahead. Sheikh Mujib's response to the postponement was to call for a general strike in the East, which Yahya countered by imposing a curfew on Dacca. Communal rioting between Biharis and Bengalis began again. On the pretext of maintaining civil order, the army moved reinforcements into the East.[17] On March 26, Yahya said that "Sheikh Mujibur Rahman's action of starting his non-cooperation movement is an act of treason. . . . He has attacked the solidarity and integrity of this country. This crime will not go unpunished." [18]

The rest of the story is well known. Army brutality was met with resistance, which brought more repression and eventually guerrilla war. Refugees streamed across the Indian border into West Bengal. The Indian government appealed to Pakistan to halt the bloodshed and to the world community to help care for the refugees. When both pleas fell on deaf ears, however, Indira Gandhi's government intervened. The ensuing war was soon over, Bangladesh was declared independent, Yahya turned over the government of what was left of Pakistan to Bhutto, and Sheikh Mujib was released from prison in the West to return to Dacca in triumph as the head of a new nation and a new government. The thirteen-year attempt of Generals Ayub and Yahya to turn political power back to civilians had been a total failure.

VII

What lessons can we draw from this case study? How can we sum up the effects of military rule in Pakistan, and, particularly, what can we say about the ability of the system to turn power back to civilians?

First, the thirteen years had seen the eclipse of the old political leaders and their replacement by men who had been martyred by the system. These new men were not generally types who were willing to make compromises with the military. *Second,* a related point, there was a dearth of moderate politicians in the country. All those who had tried to work within the military system had been disgraced. *Third,* political parties had been almost totally destroyed, with the result that contending politicians in the 1970 election were not bound by past commitments or organizational constraints. *Fourth,* the thirteen years had allowed a piling up of basic unresolved issues,

17. An excellent account of this period is David Dunbar, [pseud.] "Pakistan: The Failure of Political Negotiations," *Asian Survey,* May 1972.
18. Address to the Nation, 26 March 1971, cited by Dunbar, p. 444. Text contained in Government of Pakistan: *White Paper on the Crisis in East Pakistan,* released 5 August 1971, Appendix A, pp. 12–13.

particularly constitutional ones, so that the electoral campaign was marked by statements of positions which the military could not accept or by statements which were so vague as to be almost meaningless. No one knew exactly what could happen after the election. And *fifth,* the campaign inspired a very high level of popular mobilization, both as a protest against past policies and because elections were simply new and different kinds of events. Once such mobilization had occurred, and hopes aroused, it was impossible to send people back to their villages without results.

A military government attempting to return to the barracks, anywhere, must face similar problems. The longer it stays in power, particularly if it leaves basic economic and social issues unresolved, the harder it is to get out gracefully. The Pakistani military had gradually foreclosed all its options during thirteen years of rule, and so, when it finally said that it was turning power back to civilians, no possible solutions were acceptable. The only recourse at that point was to fight.

MILITARY RULE IN POST-WAR THAILAND, BURMA, AND INDONESIA

John F. Cady

The ending of colonial domination in Southeast Asia following World War II opened the door to a variety of revolutionary threats to its governments. In some countries, a modulation from the old to the new order was peacefully achieved, so that prewar governmental institutions persisted under much the same leadership which had been in evidence prior to 1942. Examples were the Philippines, Thailand, Malaysia, Laos, and Cambodia. In other situations, the pre-war administrative institutions and leadership were swept aside and violent controversy became almost unavoidable. Contests for control developed between rival political factions, ethnic elements, or regional groupings. Importantly present in all countries were varying degrees of hostility toward alien economic domination, both European and Asian. In most situations, a positive sense of nationality was lacking among the peoples included within the arbitrarily designated colonial boundaries which the newly independent states inherited. Traditional hostilities dating from pre-colonial times also came again to the surface, including dislike of neighbors and fear of possible Chinese intervention.

Wherever post-war constitutional forms proved unable to reconcile the various factors of conflict and distrust, the likely prospect of resort to military control was present. Order was preferable to anarchy, and the army was frequently the only disciplined agency capable of maintaining effective control. The armed forces usually posed as the defenders of national unity and independence. Because the revolutionary potentials within post-war Southeast Asia generally were too varied and complicated to be covered in a single chapter, this study will confine its comments to three states where the controlling factors were highly disparate yet capable of comparative treatment, namely Thailand, Burma, and Indonesia.

The Military in Post-war Thailand

Army rule in Thailand (pre-war Siam) dated from the 1930s. The anti-monarchial "promoters" of the *coup* of 1932 were concerned primarily with abolishing the traditional prerogatives of royal absolutism. They stopped short of destroying the monarchy as such, because its symbolic embodiment of governmental authority was too useful to discard. The established civil service administration, which had been created under royal initiative, also survived the *coup* and continued to function substantially free from political control. Thus was preserved the tradition that administrative authority derived from royal appointment and sponsorship, even though the kings were frequently absent in Europe and were powerless to intervene politically in any case. The administrative service managed to maintain a vitality and cohesion in its own right as part of the traditional pattern of governmental control.

Within Siam, the military faction of the promoter group took over control in 1937 in the person of Colonel Pibun Songgram. As the Far Eastern War approached, he assumed a pro-Japanese posture and later proceeded to utilize Japanese assistance to recover Siam's once-controlled border areas from French Cambodia and Laos, from Burma's Shan States, and from northern Malaya. Pibun's political dominance was sharply challenged during the course of the war by the Free Thai leader, Pridi Phanamyong, a lawyer rival within the Promoter group, who occupied the innocuous titular role of Regent to the youthful king who was absent in Switzerland. As the tide of war turned against the Japanese, the opposition leadership displaced Pibun in Bangkok in 1944 and developed secret connections with the Allied governmental and military intelligence agencies. The eventual Allied victory saw Pridi's Free Thai faction in control at Bangkok. Pridi enjoyed support for a time from a moderate pro-royalist faction led by the wartime ambassador to Washington, Seni Promoj, and from naval rivals of Siam's army leadership. The Allied victors required that Siam return the border territorial accretions to French and British colonial control, despite Pridi's particularly strong opposition to the reestablishment of French rule in Laos and Cambodia.

Pridi and his civilian and navy supporters were greatly embarrassed by the tragic death of the youthful King Ananda in June, 1946. The teen-age prince had been called back temporarily from his schooling in Switzerland by Regent Pridi himself to receive instructions relative to his future role as king. Ananda's death, probably by suicide, was brought on by his discouraged reaction to his pictured prospects as a powerless ruler. Because Pridi was unable to come up with a convincing explanation of the tragedy, he was blamed for it by political opponents. An army faction headed by ex-Premier Pibun Songgram, who had been freed from detention in 1946, seized control of the government in the *coup* of November, 1947, forcing Pridi and a naval officer associate to flee the country. During the initial five months of the new regime, Pibun permitted the Democrat Party leader Khuang Aphaiwong to act as Premier, before he himself took over in person in April, 1948. The trials of the three accused assassins of the king which dragged on thereafter through eight dismal years of court litigation were essentially political in character, being calculated in part to ensure Pibun's continuance in power. He withstood a *coup* staged by younger army officers in 1948.

Pibun was a consummate political actor. He capitalized on his early wartime challenge of French control of the cis-Mekong River province of Samaboury in the north by erecting an imposing Victory Monument in Bangkok. He emasculated the moderately liberal constitution of 1946, tolerated for a brief time the revised constitution of 1949, and finally cancelled all such arrangements entirely in 1951, following an abortive revolt led by naval officers. He also moved to curb arbitrarily the economic role of the resident Chinese business community in Bangkok. After 1951, he found it advantageous to share his power with two younger military officers, General Phao Sriyanon of the military police and General Sarit Thanarat, commander of the elite army unit stationed in Bangkok. After 1952, Pibun suppressed for a time all criticism of his regime, in the name of anti-Communist concern. Pibun's subsequent efforts in 1955–56 to generate a semblance of popular support by permitting a measure of free public discussion were followed by his labored move to stage new elections in 1957. By resort to wholesale bribery and intimidation tactics, Pibun and Phao managed to obtain a majority in the Assembly elected in early 1957, but their ruling days were numbered. General Sarit threw his support to the youthful critics of the election frauds and assumed control in August, 1957, forcing Pibun and Phao to flee the country. For approximately a full year thereafter, Sarit's underling General Thanom Kittikachorn exercised faltering control, while Sarit sought medical care in Europe. The latter returned to Bangkok in the fall of 1958 as army dictator, ruling under martial law decree and exercising direct personal command over both the army and the military police.

Marshall Sarit's ensuing five-year rule as dictator was more progressive economically than the tenure of Pibun had been. He followed the advice and guidance of spokesmen for the World Bank and the International Monetary Fund, who were invited to spend some time in Bangkok. His principal innovation was to relax the suppression of the activities of the Chinese business community, after persuading the latter to assign members of the ruling military elite to numerous Boards of Trustees, coupled with appropriate grants of stock. His personal avarice was matched by his ruthless repression of criticism in the press and by his famed amorous adventures with multiple wives and concubines. When he died in 1963, he left an estate valued at 140 million dollars, a considerable fraction of which had been purloined from public revenues. Whether due to the awe which he inspired or to admiration of his achievements, subsequent criticism of Sarit's predatory rule by informed Thai spokesmen was difficult to elicit. His successful methods of operation were apparently widely envied.

Power modulated after 1963 into the hands of two cooperating army leaders, allied by marriage, Marshal Thanom Kittikachorn and General Praphas Charusatien. The first took over as Premier, and the second commanded both the army and the police. Assisted by substantial American economic contributions and military construction connected with the expanding Vietnamese war, Bangkok's military dictatorship long survived under theoretical martial law. A new royally-approved constitution was finally implemented in June, 1968, and promised Parliamentary elections were held in February, 1969, despite General Praphas' strong objections. Due in part to Praphas' encouragement of Independent candidates, the Government party failed to win a clear majority. Seni Promoj's Democrat Party won a sweeping victory in the Bangkok area. On the eve of Thanom's scheduled retirement in late 1971, he staged another *coup,* abolishing both the Parliament and the constitution of 1968. Several naval leaders and Thanom's own son were elevated to the ruling elite to balance off the army opposition. Foreign Minister Thanat Khoman was retired. Thanom himself assumed the leadership of the Revolutionary Council, with General Praphas serving as Deputy head of the Council. Praphas lost out as Minister of Interior but retained his army command.

The apparent reasons for the 1971 *coup* were two-fold: (a) the lack of cooperation demonstrated by minority critics within the Parliament, and (b) the alleged threat posed by Communist agents operating along Thailand's frontiers and within the Chinese population of Bangkok. Military rejection of the 1968 constitution was probably inevitable, since General Praphas as heir apparent to Thanom was strongly opposed to democracy in any form and was more belligerently anti-Communist than Thanom himself. Moderate economic decline experienced after 1969 was also a probable factor in the reversion to absolutism, caused in part by the fall in revenue income from rice exports and in part because fewer American dollars were available as a result of winding down the Vietnam war. The undercurrent of popular protest against the *coup* was reflected in Thanom's explicit warning delivered to assembled civil service officials that the Revolutionary Council would tolerate no overt opposition or criticism. Other *coups* were virtually sure to come, probably within the ruling elite. Despite the glaring contradiction of terms, Thailand's government continued to be defined officially as a "Democracy."

Army dictatorship operated in Thailand during most of the last four decades as a substitute for royal absolutism. During the course of Sarit's five-year rule and under Marshal Thanom, Bangkok's army rulers began deliberately exploiting the traditional reverence paid to the person of the king in a deliberate effort to bolster their own ruling authority. Few of the eight or nine successive *coups* since the 1930s entailed any serious bloodshed, and the successive constitutions have left behind little substantial residue. The manifold opportunities for personal aggrandisement which have accompanied the exercise of power since 1950 were shared on a sufficiently wide scale to prevent personal rivalries from degenerating into armed confrontations. Routine governmental administration continued to function with minimal political interference; public education expanded; electric power, water services, drainage controls, and transportation needs were all effectively supplied; urban construction proceeded apace. Government regulation meanwhile kept the price of rice low for both producers and consumers by limiting sales abroad and forcing exporters to pay a differential fee to cover the variation between domestic and overseas prices. Thailand faced no extensive landlord-tenant problem. Business firms, merchants, and entertainers, mainly Chinese, shared their profits with the ruling elite and with the police. Thailand provided the unusual example of dictatorial army control combined with the encouragement of private enterprise and economic modernization. Having escaped the disruptive impact of colonial rule, the country managed to maintain intact the ancient symbols of governmental authority and the vitality of traditional social institutions, including a pervasive sense of national identity.

The Military in Burma

The advent of military rule in post-war Burma differed in a number of ways from previously described developments in Thailand. Burma's once proud military tradition died during the century of colonial rule. British rulers, after 1887, recruited no separate Burman units into the army, using only Karens, Kachins, Chins, and Indians down to 1941. None of Burma's wartime Thirty Heroes were militarily trained. Led by Thakin Aung San, they cooperated with the Japanese invaders in 1942 and contributed to hastening their exit in 1945. The Burma National or Independence Army, which emerged under Japanese sponsorship during the course of the occupation, was a non-professional organization. It ranked high in national spirit, but was in no sense a match for the trained Japanese forces or for the returning British, Indian, and African army. The nationalist allegiance accorded to the leaders of the Anti-Fascist Peoples Freedom League, especially to Aung San and the Communist Than Tun, was attributable not to military exploits but to their long-time selfless dedication to political independence. They also won the respect of Lord Louis Mountbatten, the Supreme Allied Commander for Southeast Asia, who negotiated with them as *bona fide* nationalist leaders and military allies. They withstood the negative policies of returning Governor Dorman-Smith and his pre-war Civil Service following. They later entered the government of Governor Rance in September, 1946, and negotiated Burma's independence at London in early 1947. As forthright champions of independence for Burma since the University strike of 1936, Aung San and associates won patriotic acclaim as courageous politicians rather than as military leaders.

Following Aung San's tragic assassination in July, 1947, done at the instigation of the pre-war Premier U Saw, political leadership in Burma was assumed by Thakin Nu, who had led the University strike of 1936. Nu was a devout Buddhist, an amateur playwright, who possessed no military experience whatever. As independent Burma's first Premier in early 1948, he faced the serious challenge of a Communist-instigated rebellion aided by a dissident veterans' group. During the initial stages of the rebellion, the Burma army was commanded by the Sandhurst-trained General Smith-Dun, a Karen professional.

When dissident Karen elements joined the rebel cause in early 1949, Smith-Dun was replaced by Bo Ne Win (formerly Thakin Shu Maung), one of the Thirty Heroes associated with Aung San and a long-time friend of Nu. General Ne Win continued his role as leader of the army after the rebellion subsided in 1951. He served for a brief time as Minister of War, but resigned because he had no liking for administrative duties. Down to 1958, Ne Win was content to leave governing responsibilities to Premier Nu and to his Socialist-inclined Cabinet colleagues.

Burma's army meanwhile gained both experience and morale in putting down Communist and other rebel resistance. They later became involved in attempting to counter the influence of the American-aided Kuomintang refugee elements from Communist China, who moved into the easternmost Shan border states. U.S. relations deteriorated in 1953. The AFPFL political organization, rather than the army, dominated the nationwide elections which Premier Nu staged in 1952 and 1956. Independent Burma inherited the strong British tradition of the supremacy of civilian authority over the military, but due to a variety of factors civilian rule broke down in 1958.

The failures of Nu's government were attributable mainly to administrative deficiencies, particularly its inability to follow through on the implementation of policy decisions. Civil Servants at lower levels were loath to accept such responsibility, especially when under pressure of political intimidation. Continuing disorder hampered both agricultural production and internal trade, while government efforts to carry out an overambitious program of industrial development along socialistic lines failed. Nu's personal role was confined very largely to his sponsorship of the Sixth Great Buddhist Council, which involved the construction of a costly artificial assembly cave and a gold-covered Peace Pagoda to house the alleged Buddha tooth contributed by China. Religious fervor and the exercise of personal charisma clashed with the more prosaic concerns entertained by Socialist members of the Cabinet. The latter lacked both political support and requisite financial resources. The government's classic blunder was its failure to dispose of the unstored rice crop surplus in 1956 until it was caught in the monsoon rains. Subsequent barter deals made to cover the 150,000 tons of rice consigned to the Russians for a plethora of Eastern European cement which was delivered at Rangoon immediately prior to the rainy season. Premier Nu became so discouraged with the fraudulent elections held in June, 1956, that he retired from political life for a period of eight months in a vain effort to effect reform of the party structure.

The sharp political rift which eventually developed within the AFPFL Cabinet in the spring of 1958 threatened to precipitate civil war. Nu managed in June to maintain majority support within the assembly by offering a variety of political concessions to minority factions. But his Parliamentary backing ebbed away in August prior to the scheduled calling of general elections. Political tensions reached the breaking point in September with elements of the military police pitted against the army. It was finally with the consent of both factions of the AFPFL that General Ne Win and the army were asked to take over in a caretaker capacity for a limited six months period (later extended to sixteen) to pacify the country and to arrange new elections. The ensuing period of Caretaker rule cancelled the unrealistically planned economic development program, cleared up squatter hutments in Rangoon, collected accumulated garbage, and got rid of droves of pariah dogs. Officers of the army took over numerous duties relating to the distribution of food and other consumer goods, in addition to their policing functions.

True to his promise of 1958, General Ne Win staged new elections in early 1960, which unexpectedly brought Nu back to power. Taking advantage of popular dislike of army rule, Nu promised increasing measures of autonomy for minority groups and proposed that Burma would become a Buddhist state, while affording protection for other faiths. He thus corralled a large but unmanageable Parliamentary majority. During the ensuing two years, Nu permitted a number of the army's operational agencies to continue their caretaker functions, while his own administration floundered badly without the help of former Cabinet member opponents who had failed election. Although the general agricultural situation tended to improve of its own momentum, Buddhist partisans came to blows with Muslim and Christian groups over the Buddhist-state issue, garbage and ownerless dogs reappeared, and positive direction was lacking. Premier Nu, highly discouraged, talked of political retirement, thereby arousing the ambitions of rival leaders anxious to take over, especially those within the army who were waiting impatiently off stage. Particularly objectionable to General Ne Win were Nu's efforts to carry out his promises of regional autonomy.

The eventual army coup of March 2, 1962, occurred in conjunction with Nu's convening of a conference of Shan leaders at Rangoon to work out the details of the future status of the Shan states located in the eastern plateau region of Burma. The army intervened by arresting all participants in the conference along with members of the Cabinet, the Supreme Court, and all potential political opponents. The constitution was abolished and power was assumed by a self-appointed Revolutionary Council composed mainly of army officers. The move was patterned in conformity with General Sarit's action at Bangkok in 1958, but the differences were substantial. Whereas the Thai dictator could count on the continued routine functioning of an experienced apolitical civil service and was willing to accept the guidance of World Bank and other competent advisors, Ne Win's military Council was entirely on its own administratively and hopelessly befuddled about what to do. When the best available leadership of both factions of the divided AFPFL refused to assume Cabinet responsibilities unless the constitution was restored, the Revolutionary Council eventually accepted the volunteered services of several Marxist leaders of the National Unity Front coalition. Once in control, the NUF group proceeded to formulate the doctrinaire *Burmese Way to Socialism,* which was designed theoretically to end private capitalist exploitation of workers and peasants by establishing state control over production facilities and trade. As soon as workers and peasants could learn to appreciate the advantages of the new order and to accept their responsibilities for its proper functioning, political power would be vested in a hierarchy of Soviet-type councils under Democratic-centralist control. Until such time, the self-appointed Revolutionary Council would exercise plenary control, while recruiting additional personnel through probationary procedures.

Although Ne Win was in no sense a convinced Marxist and continued his army operations as before against the Communist rebels in central Burma, his government embarked on a fumbling attempt to establish a Socialist paradise in which man no longer exploited fellowman. Indian and Chinese entrepreneurs were the major initial targets, many being forced to leave the country without their savings. Capitalist opponents of the regime were prevented from functioning economically, as army personnel took over all essential services. Ex-employers were required to continue paying wages to discharged workers and were also forced to pay taxes on non-existent income in proportion to pre-1962 schedules. Government-run Peoples Stores were eventually assigned the distribution of all rationed consumer goods in short supply. Miserably paid store managers operated profitably through the back door with the inevitable black marketeers in order to make ends meet. On the agricultural front, peasants repaid only a fraction of government-provided loans. They also surrendered only a minimal proportion of their surplus rice to Government purchasing agencies in order to get their cloth ration tickets, while hoarding the rest for family consumption and barter purposes. When available rice supplies for export fell to a small fraction of the anticipated volume, government purchasing agencies responded by curtailing basic imports for both consumer and capital needs. Meanwhile, the favored army leadership and their civilian friends enjoyed continued monopoly access to houses, motor cars, electrical appliances, and other luxury items including moth balls and golf equipment. Growing discontent over the downward spiral of living standards was permitted no overt expression. Protesting Buddhist monks as well as University students were shot down by army rifle fire, and after a few such instances none dared take liberties with the General's ungovernable temper.

Whereas army dictatorship in Thailand became a substitute for royal absolutism, Burma's Revolutionary Council attempted to fill in for the nonexistent Communist Party hierarchy. Rangoon authorities introduced Marxist propaganda in the schools and enlisted Soviet assistance in both educational and economic development. Ne Win's government, like that of its predecessor under Nu, carefully avoided giving offense to neighboring China, but it refused to become subservient. When the Red Guard agitators infiltrated the Rangoon schools in June, 1967, Ne Win suppressed the effort, shipped home all Chinese advising experts, and cancelled the Peking aid program. Eastern European observers in Rangoon reportedly expressed concern over the embarrassing fact that the steadily deteriorating economic system operating in Burma was called "Socialism." Generally speaking, power considerations relating to Burma's foreign policy were equally important with ideology. American government agencies, for example, continued the sale of small arms and ammunition at discount rates to Ne Win's Nationalist-Communist regime throughout the first ten years of Ne Win's army dictatorship. Well understood was the fact that Burma's rulers would not permit any infringement on the country's sovereignty.

The Army's Role in Indonesia

The Indonesian army of post-independence times was made up of several highly disparate fragments. Approximately half of the 65,000 men of the Netherlands colonial army, consisting mainly of Christian Indonesians and Eurasians, were disbanded after the departure of the Dutch. Some units were retained because of their particular competence and others from concern lest their abrupt dismissal might precipitate difficulties in various areas of the Republic. A portion of the demobilized fraction of the colonial army actually staged a short-lived *coup* in January, 1950, under the leadership of a renegade Dutchman named Westerling. The guerrilla-type anti-colonial army began to fall apart once the common Dutch enemy was eliminated. Some officers of the Japanese-sponsored Peta army continued to entertain the view that their patriotic army should rightfully assume an active political role alongside the allegedly selfish political party organizations. A Western-educated and less professionally minded officer group at first discounted the political legitimacy of the army politicians, although some of them later also came to doubt the feasibility of democratic rule. A third fraction, who obtained their professional training in the United States and elsewhere in the West came to prefer a smaller but better disciplined and technically proficient military establishment. Some of the latter wanted to eliminate all unprofessional guerrilla forces who might be tempted to regard the army as a career.

Further fragmentation developed in time on the basis of regional origins and ethnic identification. Recruits operating within their own neighborhoods as guerrilla bands exploited the obvious advantage of close relations with their own people. Officers who were obliged during the course of the challenge of Dutch rule to assume political and administrative roles were loath to abdicate such prerogatives to civilian officials after independence was achieved. The Republican constitution of 1949 explicitly denied to all the constituent states of the contemplated Federation the right to maintain armies of their own, but this ruling was not strictly enforceable.

When a move was later made to nationalize all Dutch-owned properties at the outset of the Irian crisis in 1957, local military leaders took over lucrative economic roles as perquisites of power. In favored spots, local commanders were able to market exportable surpluses directly to Singapore and elsewhere to their own personal and regional advantage. Corruption also reached fantastic proportions in the heavy tolls levied during the 1960s on foreign concession explorers for oil and tin. When reentry into civilian life proved difficult, some of the ex-guerrilla units turned to banditry. In parts of western Java and south Sulawesi, fanatical proponents of a theocratic Islamic state (Darul Islam) maintained centers of resistance for a number of years. Thus the military residue of Indonesia's independence struggle proved far from orderly and homogeneous.

The first important political initiative was taken by the Indonesian army in October, 1952, when the economy-minded Wilopo Cabinet proposed to dismiss a substantial proportion of the guerrilla contingents in both the army and the military police. When several pro-Sukarno Assemblymen moved to block the modernization effort, a group of army leaders staged a march on the Assembly. They demanded in vain the dismissal of the Assembly, and the leader of the group, General Nasution, was temporarily removed from his command. Sharp divisions which developed within the army ranks were later combined with disagreements between the Cabinet and the Assembly to force the resignation of the able Wilopo Cabinet in August of 1953. At this point the political influence of the army was clearly subordinate to the rising political power of President Sukarno himself. But the pro-Sukarno successor Cabinet of Sastroamidjojo came to grief some two years later (June, 1955) when it tried to install a new army Chief of Staff who was not acceptable to the senior officers. Political partisanship rose sharply as a result of the elections of late 1955, which produced a virtual impasse between the Sukarno-led faction and the religiously oriented Masjumi grouping.

It was in 1956 that the army and the President eventually found common ground in their joint opposition to political-party government derived from the previous year's election. The same opposition to party rule was shared by the Javanese Communist faction, which was being denied a place in the coalition government despite its impressive election performance in 1955. Contributing to growing disillusionment with democratic rule was the 1956 spectacle of constant party bickering, administrative corruption and inefficiency, plus mounting monetary inflation and the government's failure to come to grips with basic economic problems. The traditional Javanese ideal of government by consensus was clearly unattainable in the face of such party rivalry.

The transition in 1957–1958 to Sukarno's Guided Democracy pattern was accompanied by the assumption on the part of the military authorities everywhere of increasing independence from Cabinet control. Semi-autonomous military commanders in three separated provinces of Sumatra, who were short of funds to meet payrolls, openly supplemented their local revenues by selling exportable commodities directly to Singapore. These efforts in time blossomed into short-lived rebellions backed politically by Masjumi-Socialist malcontents, setting an example for similar efforts in Eastern Indonesia later in 1957. The central army command managed to recover control at the time by juggling command assignments. In the end the Outer Island commanders accepted the authority of the army chiefs in return for the exercise of a considerable degree of autonomy in their local administrative and economic roles. Meanwhile, political control at the center was monopolized by Javanese parties, including Sukarno's Nationalists (PNI), the conservative Nahdatul Ulama, and the Communist propagandists who were active at the village levels.

As finally developed, Guided Democracy became a kind of political balancing act in an attempt to mitigate prevailing political, social, and cultural rifts. Capitalizing on his charismatic showmanship, the President undertook to revive and sustain revolutionary fervor and nationalist spirit at a feverish pitch. Basic problems of declining economic production, unfavorable trade balances, and runaway price inflation were pushed aside as secondary concerns. Sukarno permitted the Communist Party to continue active, while eventually outlawing all other groups. He would promote revolutionary fervor against democratic institutions, but he denied the PKI any role in the actual government.

Army participation in the government down to 1962 centered on the strenuous demand that West Irian be surrendered by the Dutch, a movement accompanied by the nationalization of virtually all Dutch properties within Indonesia proper. Once the West Irian issue was resolved, Guided Democracy shifted its major focus to challenge the British creation of Malaysia in 1963. These strenuous anti-Western moves involved Djakarta's repudiation of its United Nations membership, its acquisition of substantial military hardware from the U.S.S.R., and the increasingly close affiliation of the PKI with the viewpoint and policy concerns of Communist China. Meanwhile, inflationary prices within the islands mounted to intolerable levels, as exports declined and domestic economic problems were entirely neglected.

The bizarre alliance between President Sukarno, the conservative Nahdatul Ulama, the Communist Party, and the army was too artificial to last indefinitely. The army leadership, for example, was much less inclined to become involved in the Malaysian Confrontation than were the Communists and their cooperators within the Air Force. Lower officers of both military services were jealous of the brass. The PKI meanwhile sought in vain for permission to develop its own armed auxiliary force for use in Borneo along with repeated demands for active Communist participation in the Djakarta Cabinet.

The alliance finally fell apart during the late summer of 1965 when Sukarno suffered a prolonged illness. A *coup* attempt took place in the early morning of October 1, when malcontent elements within the lower levels of the army command, backed by elements in the Air Force, attempted to assassinate all top army leadership. Charging that the army itself had been about to stage its own *coup* against the ailing President, the assassins solicited Sukarno's personal support and also enlisted the cooperation of Communist Party leaders, who were not averse to seeing their major political enemies destroyed. The only two high army officers who escaped death were General Nasution, who suffered an ankle injury while fleeing his house, and General Suharto, who was not included among the list of victims presumably because he was reportedly a friend of Sukarno.

The ensuing counter action by the army was led by Suharto, whose forces promptly captured the airport where the rebel elements had assembled. Once the rebellion was brought under control, the army blamed the Communist Party for the entire affair. The tactful Suharto treated the ailing President with personal deference, so as not to provoke a hostile reaction from his many admiring followers in Java, but the army assumed complete control of the government. The trial and execution of rebel military leaders was followed several months later by a gruesome pogrom, deliberately instigated by the army, against Communist party leaders and their following in Central and Eastern Java. The victims numbered several hundred thousand. Sukarno's repeated efforts to recover control of the government were systematically blocked by Suharto, but the national hero President was permitted to live out his time until he died from cumulative ailments. Once in firm control, General Suharto invited into his emergency Cabinet a number of able civilian leaders, some of whom had been associated with the anti-Sukarno movements of 1956–1957. The Confrontation of Malaysia tapered off and was finally abandoned; Indonesia later reentered the United Nations and reversed its pro-Communist orientation.

The recovery tasks which the Suharto government faced were enormously difficult. Monetary inflation had to be brought under control and foreign debt obligations renegotiated. The latter task was accomplished with Western help, especially from the Dutch, who were also permitted to reoccupy a number of their colonial economic holdings, many of them near total collapse. Additional credits were eventually forthcoming and development efforts got under way. Abuses of power on the part of army personnel continued in many areas of control, especially in outlying regions of the republic, but full support was accorded by the central government to proposed reforms essential to economic recovery.

Although the political difficulties were formidable, the Djakarta authorities moved in time toward the restoration of a modified form of constitutional government. The proposed constitution was qualifiedly democratic, but stressed functional representation rather than regional or party groupings. Despite the great effort and cost involved, general elections were held on schedule in July, 1971. The military rulers took few chances, insisting that the army itself should be accorded a full 100 seats in the new 470-member Assembly, and vetoing active political roles for *Masjumi* Party leaders. But relevant issues were discussed and the government-sponsored Sekber Golkar slate won a substantial majority at the polls. The two governing Councils which were formulated following the elections, one to handle policy decisions and the other the day-to-day administration, both contained a majority of army members. The same able civilian leadership continued in office, and their economic recovery policies are firmly supported.

The essential differences between army rule in Burma and in Indonesia related to the broader and more intelligent outlook of the Djakarta leadership, who were enlisted for the performance of essential tasks. Indonesia's Foreign Minister Malik was later elected to the post of President of the United Nations General Assembly, whereas a person of U Thant's comparable capacity and independence would be *persona non grata* at Rangoon. Faced by economic crisis, the Djakarta authorities permitted the resumption of alien operation of many important facilities confiscated under Guided Democracy and were willing to give assurances needed to attract new investment. The guidelines were set not only by older creditors but also in accordance with advice given by the World Bank and International Monetary Fund experts. Bribe-taking on a local scale continued, particularly in such areas as the granting of concessions for undersea oil and tin prospecting, and full economic recovery of Indonesia still had a long way to go.

Comparisons

Rangoon's do-it-yourself tactics under the *Burmese Way to Socialism*, by comparison with neighboring countries, ran completely aground, despite the fact that Burma's physical and human resources were relatively speaking fully equal to those of Indonesia and Thailand. General Ne Win's xenophobic dictatorship became committed to a doctrinaire Marxist program which smothered all dissent and was completely unable to elicit popular cooperation or effective administration of planned development schemes.

This comparative examination of army rule in three Southeast Asian countries demonstrates among other things that military rule can develop under a variety of circumstances. The quality of the leadership, the ends pursued, and the conditioning circumstances have clearly produced highly disparate results. The three situations also demonstrate that American efforts to promote democratic governmental practices and counter Soviet efforts to promote Communism can both be futile if pushed without a sensitive awareness of the situations involved. The revolutionary developments which have taken place in such differing environments cannot be adequately explained in Democratic or Marxist economic terms, but must be related to the entire spectrum of social and cultural factors contributing to unrest. Much more could be affirmed along similar lines with respect to revolution and army rule in post-war Vietnam, granted the eventual achievement of adequate perspective and free access to essential sources of relevant information.

Few will question Lord Acton's famous dictum that "Power corrupts, and absolute power corrupts absolutely." Military rule can serve as a stopgap in crisis situations, but it suffers corrosion over the long term. The several alternatives to the dismal prospect of the inexorable abuse of arbitrary power are suggested in the experiences of the three countries here described. The ultranationalist concern which characterized Burma's post-war policies, aimed against all foreign and minority elements, has exacted a forbiddingly heavy price which has been aggravated by a decade of army dictatorship. Nationalism clearly reached the point of diminishing returns in Burma's xenophobia. The road back would appear to lie in a return to some form of constitutional government, which would place limits on government powers and also define the rights of citizens in broadly human terms.

It can be anticipated that public opinion in Thailand will eventually demand a halt to the abuse of power by the ruling military elite. Such an effort can be pursued within the context of traditional social and cultural values and accepted national identity, including continuing respect for the monarchy. Bangkok in time will have to return to some form of constitutional rule if economic progress is to continue and the threat of revolution averted. Although army rule in Indonesia may continue for some time into the future as a guarantee of stability for the widely extended republic, persecution of the Chinese business community has already been contained, able men have taken over crucial places within the government, and relations with the outside world both economically and politically have been recast on a rational basis.

A final comment relates to the role which any genuinely friendly nation can perform constructively in such situations. Excessive concern for the arbitrary dichotomy of the Cold War confrontation will have to be curbed.

Particular situations in Southeast Asia as elsewhere in the world cannot be viewed as mere pawns in the world-power chess game. Neither Western-type democracy nor anti-Communist military dictatorship are likely to meet particular needs; nor are the peoples concerned likely to choose Communist-type regimentation for any long-run period. If constructive progress in terms of meeting human needs is to become possible, outside powers as well as the peoples concerned will have to take into account the wisdom which experience and full understanding of the problems can provide. The outlook need not be one of discouragement and despair.

THE SOLDIER AS CIVIL BUREAUCRAT IN A DEVELOPING COUNTRY——KOREA

Paul S. Kim

After their successful coup in 1961, Korea's military leaders committed themselves firmly to the nation's socioeconomic development and security. Impatient with the apparent procrastination, corruption, and inertia of their civilian predecessors, they promised economic progress and movement from a transitional to an industrial society. Moreover, capitalizing on the Koreans' fear of Communist aggression from the North, they promised to establish a strong military force capable of maintaining national security.

During the formation of their regime, the military leaders moved to consolidate their power by bringing many of their fellow soldiers into the civil bureaucracy. By 1966, nearly fourteen percent of the top-level civil service positions (231 out of 1,640)[2] were occupied by ex-soldiers, most

1. The author wishes to express his gratitude to the Gannon College Faculty Senate for providing a research grant which enabled him to travel to Korea to gather materials for this article. Among many others, he especially thanks Professor Dong Suh Bark of Seoul National University; Professor Woon Tai Kim, Dean of the Graduate School of Public Administration, Seoul National University; Professor Moon Young Lee, of Korea University; and Mr. Jae Koo Ha, Director of Parliamentary Procedure, National Assembly, Republic of Korea, for their stimulating conversation and generous assistance in his research. He is also grateful to Professor George Welch of Gannon College and to Albert Somit, Executive Vice-President, Buffalo University Center, State University of New York, both of whom read the manuscript and provided valuable criticisms. The opinions and comments in this article are solely those of the author, and he assumes full responsibility for them.
2. Gong Mu Won Je Do ["The public service system"] (Seoul: Graduate School of Public Administration, Seoul National University, 1966), pp. 55–64.

of these former officers who had been brought into the civil service through special recruitment procedures at relatively early ages, and now mostly in their thirties and forties. Together with the former soldiers in the legislature and in the government corporations, the military men in the civil bureaucracy have had a strong role in the process of Korea's development.

The military leaders' concept of development, an admixture of ideas ranging from common sense to ultramodern economic theory, stresses socioeconomic development rather than political development. In their view, political democracy will inevitably follow if Korea achieves sufficient socioeconomic development. Their basic concept of development has three major aspects: the socioeconomic climate; the agents of reform; and the strategies of reform.[3] They believe that the mere transplantation of Western ideologies and methods of development to a developing country like Korea cannot succeed unless the socioeconomic climate, determined by factors such as education, religion, political culture, political system, etc., is conducive to the country's growth.

A strong educational system, they believe, is essential to socioeconomic growth, and ease of communication and transportation will foster the development of innovative ideas.

Without dynamic political leaders able to provide stable yet effective administration, however, economic growth is almost impossible in a developing country. The socioeconomic climate in such a country depends largely on the agents of reform, those who must not only initiate development concepts, but also implement them under the political leadership; this task falls to the civil servants, especially those at higher levels. A poorly educated, ill-trained civil servant who cannot cope with social changes but tends to long for the "good old days" may retard any solution of present problems or action which might lead to future growth. On the other hand, a progressive public servant who sees social changes as opportunities for experimentation and exploration is apt to work in the present to assure future gains. A developing country must be able to recruit an administrative elite of innovators who, in conjunction with its political leaders, will determine the strategy of reform—what reforms will be undertaken, by whom, when, and for what ends. The choice of strategies will be limited by the socioeconomic climate and the time available, but will also be affected by the personal background of the agents of reform, i.e., their age, education, religion, previous experience, family life, etc. Thus, in the military leaders' view, the establishment of an effective po-

---

3. There is no single official account of Korea's recent socioeconomic development, but substantial evidence of its economic development can be found in the reports of the Economic Planning Board of the Republic of Korea. Jae Ho Kihl, a close associate of President Park and the architect of the constitutional amendment providing for a third presidential term, treats Korea's growth in his book *Usumi Hwaljack Piltai* (Seoul, 1969).

litical-administrative system and the recruitment of able administrators are the keys to success in Korea's socioeconomic development.

This article will analyze the concepts and practices of development followed by the Korean civil bureaucracy under military leadership during the years 1961 to 1972, with particular attention to the administrative system, the recruitment of civil servants, and the administrative process.

I. How the Soldiers Became Civil Authorities

There are two accounts, one official and one unofficial, of how the military leaders came to civil power in Korea. The first view is that promulgated by the former military leaders who have been active in the Third Republic; the second is that held by the leading Korean intellectuals and by some substantial part of the population at large.

According to the official version,[4] the governmental system of the Second Republic (1960–1961), a decentralized political and administrative system based on the democratic principles of bi-cameralism, parliamentarism, and local autonomy, had nearly every conceivable defect of structure. It increased the difficulty of maintaining nationwide policies for economic and political progress; the decentralized structure gave too much power to local officials who were provincial and short-sighted in their view of national policies; the bicameral system not only wasted time and money, but also allowed one chamber to avoid its responsibility by sending ill-considered legislation to the other; and the multiparty system in the legislature resulted in an unstable cabinet system.

According to the military leaders, these structural shortcomings of the political system entailed social evils. Nepotism became widespread, and the administration grew less efficient each day. The government could not even regulate the price of consumer goods; between December 1960, and April 1961, the price of rice jumped by sixty percent, while coal and oil costs rose more than twenty-three percent. In the same period the crime rate more than doubled, but the arrest rate dropped to sixty-eight percent (it had been ninety percent in the First Republic, 1948–1960).[5] The military leaders, viewing the ineffective operations of this system in the Second Republic, felt that the result would be massive political, administrative, and economic chaos. The state of social, economic, and political upheaval, combined with other factors, led to the overthrow of the Second Republic by a military coup on May 16, 1961.

This official version is saturated with the ardor of the military leaders'

---

4. Kihl, *Usumi Hwaljack Piltai,* pp. 116–121; Pak Joong, "Park Chung-Hee—General under Limelight," *Korea Journal* 1 (September 1961): 12–14.
5. *United Nations Economic Bulletin for Asia and the Far East* 11, 3 (1960–1963): 67. See also *Haptong Yongam, 1961* (Seoul, 1962), pp. 152–53.

crusade to save their nation from its social ills. According to this view, the leaders of the military coup sincerely believed that a "surgical operation to eliminate corruption and eradicate other social evils" would benefit the entire nation, and that they had acted solely out of patriotic motives.

Yet, according to the unofficial view,[6] the military leaders' representation of the civil bureaucrats of the Second Republic as inefficient, disorganized, and sanguine was a deliberate exaggeration. Indeed, corruption and financial irregularities among public servants in 1960–1961 were no worse than they turned out to be in the late 1960s under the military regime, and many of the bureaucrats of the Second Republic, whom the military leaders had called corrupt and inefficient, continued to hold important positions even after the coup. The adherents of this latter view concede that there were political disorders in the Second Republic, in the form of strikes and demonstrations, but regard those incidents as simple manifestations of popular political feeling. Those who accept the unofficial account contend that the military leaders overthrew the popular government by force to attain their own ends, and that the regime and laws they established were merely the instruments for achieving those ends.

After the coup of 1961, the military leaders, whatever their grounds for seizing power, called themselves the Supreme Council for National Reconstruction and ruled as a junta until they formed the Third Republic on October 15, 1963. The junta's leader, General Chung Hee Park, became president, while many other ex-soldiers exchanged their uniforms for civilian clothing and became civil bureaucrats, legislators, and government officials.

II. The Political-Administrative Structure for Development

The formal structure of the Third Republic (1963–1972) was not very different either from that of a Western democracy or from that of the First Republic (1948–1960),[7] but differed markedly from that of the Second Republic, which was characterized by decentralization and parliamentarism. It was a presidential system with some modifications, more nearly akin to the political structure of the Fifth Republic in France than to that of the

6. The author has received this account from many people during his visits to Korea in 1968, 1969, and 1971, but wishes to respect their request to remain anonymous.
7. See the Constitution of the Republic of Korea, as amended on December 26, 1962, and on October 21, 1969. The best accounts in English of the governmental structure of the Third Republic are: Se-Jin Kim, *The Politics of Military Revolution in Korea* (Chapel Hill, 1971); John Kie-Chiang Oh, *Korea: Democracy on Trial* (Ithaca, 1968); Kyung Cho Chung, *Korea: The Third Republic* (New York, 1971); Se-Jin Kim and Chang Hyun Cho, eds., *Government and Politics of Korea* (Silver Spring, Md., 1972).

United States. It had three branches—legislative, executive, and judicial—
and was originally designed with a set of checks and balances like that of
the United States. After the constitutional amendment of 1970, however,
it became more nearly a parliamentary system, and a member of the National Assembly could then become a cabinet member while keeping his
seat in the legislative body.

The governmental structure established under the Constitution of the
Fourth Republic (November, 1972—) is characterized by the centralization of power in the hands of the president. Full legislative authority resides
in the National Assembly, which also has the power to make appropriations,
to ratify treaties and declare war, to impeach officials, and to recommend
the dismissal of the prime minister and members of the State Council. Its
organization into standing committees and its adherence to parliamentary
procedure are generally the same as in Western countries. Yet the Constitution of the Fourth Republic has made some significant and substantive
changes whose total effect is to weaken the legislative branch of the government: the term of office of an assemblyman has been lengthened from four
to six years; one-third of the 219 members of the National Assembly are
appointed by the president, with the consent of the National Conference
of Unification, while the other two-thirds are directly elected by a secret
ballot of all adult citizens in their constituencies; the National Assembly no
longer has the power to review matters of public administration, although
it can still appropriate or withhold fiscal support for the civil service; the
members of the Assembly are prohibited from making any remark derogatory
to the president; and the Speaker of the House has been given strong powers
to apply disciplinary sanctions against any member who obstructs parliamentary procedure through a filibuster or demonstration. In short, the new
Constitution makes the National Assembly little more than an instrument
for ratifying the policies of the executive branch. Moreover, the election of
27 February 1973 has solidified President Park's control over the Assembly:
his Democratic Republican party won seventy-three of the 146 seats at
stake, while the opposition New Democratic party won fifty-two, the
Democratic Unification party two, and independent candidates nineteen.

The Constitution of the Fourth Republic also provides for a judiciary
consisting of a Supreme Court and appellate, district, and family courts,
whose jurisdictions are substantially the same as they were under the Third
Republic. The Supreme Court justices are appointed by the president for
ten-year terms on the nomination of the Chief Justice, while the Chief
Justice is appointed by the president with the consent of the National Assembly. Yet a drastic change in the function of the Supreme Court is clear
from the fact that it no longer has the power of judicial review. Any question of the constitutionality of legislation enacted by the National Assembly
is to be decided by a Constitution Committee, whose nine members are
appointed by the president. Furthermore, the same body will have the

ultimate decision on questions of impeachment or of the dissolution of political organizations. Thus, the Supreme Court cannot be expected to function under the new Constitution as an effective and independent branch of the government.

According to the new Constitution, the president is responsible for all the activities of the executive branch; he holds office for a six-year term, but there is no limit to the number of terms he may serve. He is specifically charged (Article 43) with working toward the reunification of Korea; he has the power to dissolve the National Assembly (Article 59), but is not subject to a vote of confidence by the Assembly; and, by virtue of his power to appoint the members of the Constitution Committee and one-third of the members of the National Assembly, he has substantial control over both the legislative and the judiciary branches of the government. He also enjoys vast powers as the chief of state, chief administrator of the executive branch, chief diplomat, and commander-in-chief of the armed forces. His political power is enhanced by his position as the chief lawmaker, leader of his party, and spokesman for his people. Thus, the Korean president holds power in his own country to an extent matched by few other men in the world.

Among the distinctive features of the new Constitution was its creation of the National Conference for Unification, which has the power to choose the president, approve the president's appointment of members of the National Assembly, ratify constitutional proposals originated by the National Assembly, and deliberate major policy questions pertaining to Korea's reunification. On December 23, 1972, the 2,359 delegates to the National Conference for Unification selected incumbent President Chung Hee Park to serve for a six-year term as president of the Republic of Korea.

The presidential Executive Office, which has not been greatly changed under the Constitution of the Fourth Republic, consists of the Chunghadae staff (comparable to the White House staff in the United States), the State Council (cabinet), the National Security Council, the Economy-Science Consideration Council, and the Central Intelligence Agency.[8] The members of the Chunghadae staff, the most important organization in the presidential Executive Office, are responsible directly to the president. Often long-time associates of the president, and absolutely loyal to him, they are appointed by him and serve completely at his pleasure. The Chunghadae staff is divided into two units concerned with political affairs and administrative affairs. The latter unit is subdivided into groups dealing with general affairs, public information, petitions, intelligence, and protocol. The political affairs unit includes seven presidential assistants, the chief one responsible directly to the chief of the Chunghadae staff. Thus, after the president

8. *Choi Ko Kwal Li* ["Top management"] (Seoul: Graduate School of Public Administration, Seoul National University, 1965), pp. 9–51.

himself, the chief of the Chunghadae staff is the most powerful individual in the Korean government.

The major organizations within the executive branch are the ministries, currently numbering thirteen. These deal with foreign affairs, domestic affairs, finance, justice, defense, agriculture and forestry, education, reconstruction, commerce and industry, transportation, health and social affairs, communications, and culture and information. They are supplemented by two boards, four offices, and twelve government agencies. Each board (*won*) or office (*che*) is treated as a regular ministry, and its head holds cabinet rank. Each ministry is headed by a minister who is under the formal jurisdiction of the prime minister, and who is directly responsible to the president for the satisfactory completion of routine duties and the effective implementation of established policy. The prime minister, the chief assistant in the State Council, is appointed by the president. Under the constitution, the President also has the authority to appoint a deputy prime minister; this position is now held ex officio by the minister of the Economic Planning Board. The president has formal jurisdiction over the heads of these executive bodies through the prime minister, but in practice often bypasses the prime minister and exercises his control directly. In formal structure, the ministries are relatively simple, with few variations. Under each minister are a number of offices, boards, services, and bureaus, each responsible for a particular specialized area. These units are in turn broken up into sections, which are further subdivided into groups. The various ministries can also be divided according to line and staff functions.

Because the military leaders desired to retain close control over local authorities, local administration has under the Third Republic been placed under the direct supervision of the central government.[9] The military leaders have never given the principle of local autonomy any recognition in their political philosophy. Korea is a unitary state, mapped off into artificially determined districts for the more convenient performance of governmental functions. It has nine provinces, but the provincial governments are merely administrative subdivisions of the executive branch of the national government. The provincial governments are, in turn, subdivided into counties (*kun*) or cities, and villages (*myun*) or towns. The nine provincial governors, along with the mayors of the special cities of Seoul and Pusan, are appointed by the president and responsible solely to him. As of 1971, Korea had 139 counties, 30 cities, 1,382 villages, and 91 towns.[10]

Together with its reforms of the political structure, the military regime introduced modern techniques of public management to enhance govern-

9. Kim and Cho, *Government and Politics of Korea,* pp. 91–126. A good treatment of the subject in Korean is Byung Gun Kang, *Hankook Chibang Haengjong* ["Korean local administration"] (Seoul, 1965).

10. Kim and Cho, *Government and Politics of Korea,* pp. 110–112.

mental efficiency. Western methods such as the planning and programming system, report control, work measurement and simplification, briefing by charts, and record management were adopted.[11] Also, two administrative agencies, the Administrative Management Bureau (in the Ministry of Government Administration) and the Administrative Improvement Research Commission were set up to study potential improvements in public management and organization, and to recommend to the appropriate authorities measures to overcome existing defects. The implementation of governmental policies was made more effective through a variety of means, including incentives and education, compromise and coercion. Korea's military leaders introduced many new structural reform plans, and succeeded in fulfilling virtually all of them.

The effective operation and dynamism of government agencies under the military leadership of the Third Republic created a relatively stable regime which fostered the development of Korea's economy in several important ways.[12] First, the combination of cheap skilled labor and attractive investment laws in a stable nation encouraged much-needed foreign investments and loans; second, the Korean people themselves developed the confidence to make long-range plans and invest their own money and time in various enterprises; third, the country's political and social stability induced many young Korean scientists and engineers, educated in the West, to return home, thus forming a pool of scientific knowledge.

Despite Korea's economic gains and its increased political stability, its military leaders took a negative view of public participation in the decision-making process. They saw the democratic process, embodied in public debates, political compromises, and citizen participation, as wasteful, corrupt, hypocritical, and above all inefficient. They therefore distrusted the politicians, intellectuals, college students, and others who so vociferously espoused the cause of democracy. In December 1971 the president proclaimed a state of national emergency, justifying his action as necessary preparation for negotiations with North Korea. In the spring of 1972, he assumed emergency powers to control the nation's economy. On October 17, 1972, he declared martial law throughout South Korea, "suspended part of the Constitution, dissolved the National Assembly, and suspended

11. Suk Choon Cho, "Korean Experience with Administrative Reforms since her Independence," *Hang Jong Ron Chong* ["Korean journal of public administration"] 7 (1969): 287–300.
12. On Korea's recent economic gains, see the *New York Times,* 18 April 1971, section 12 (an advertisement); William P. Bundy's article in *Newsweek,* 11 January 1971, is progovernment in its outlook; Seung-Hee Kim, "Economic Development of South Korea," in Kim and Cho, *Government and Politics of Korea,* pp. 148–75. In Korean, see Hahn-Been Lee, "Balchon Haeng Jong Ron ui Riron Jongai" ["The theoretical context of development Administration"], *Hankook Haeng Jong Hak Bo* ["Korean public administration review"] 2 (1968): 287–98.

all political activities. Press censorship was also imposed, and all universities and colleges were closed" by the military regime.[13]

Five weeks later, on November 21, 1972, came an event unique in the nation's political history, the endorsement in a national referendum of the Constitution of the Fourth Republic, which provided for the president's election not by the people, but by the National Conference for Unification; removed the former limit on the number of terms a president may serve; created a Constitution Committee; and established the practice of electing members of the National Assembly from multi-member districts. In President Park's view, because the irresponsible political parties had already lost their sense of a national mission, these measures were needed to bring South Korea into touch with present-day realities, and to support his active pursuit of negotiations with North Korea. In short, he argued, he took these drastic steps to save the nation from political confusion and move it toward unification. But his critics believed that the 55-year-old president, a former general who had taken power in a coup d'etat in 1961, wanted to remain in power for life. Under the old constitution he would have had to step down in 1976, when his third four-year term expired. These critics further contended that the measures taken by President Park perhaps reflected the wishes of the former soldiers serving in the civil bureaucracy, who hoped to perpetuate their tenure in office in the same manner as the president himself.

III. The Civil Bureaucracy and the Bureaucrats: Agents of Reform[14]

The military leaders were aware that if an important task is assigned to someone unable to handle it, or if a public servant is dissatisfied, the work suffers, and the government may never reach its stated objectives. They saw that the success or failure of the Third Republic would to a large extent depend on the strength of its personnel. Consequently, they undertook to reform the civil service system, and particularly the procedures for recruiting public bureaucrats.[15]

The Korean civil service comprises two branches: the regular service,

13. *New York Times,* 22 October 1972.
14. Useful books on the Korean public service system include: Tong Suh Bark, *Hankook Kwal Ryojeido ui Ryuksha Jek Jonkai* ["A historical development of the bureaucracy in Korea"] (Seoul, 1961); Tong Suh Bark, *Insa Haeng Jong Ron* ["Public personnel administration"], rev. ed. (Seoul, 1965).
15. *Kong Mu Won Bop* ["Public service Law"], Article 2. The annual report *Hanguk ui Insa Haeng Jong* ["Korea's public personnel administration"], produced by the Ministry of Government Administration, is very useful with regard both to the public administration organizations and to the procedures of recruitment, examination, appointment, classification, promotion, etc.

divided into five grades, and the special service. Grade 1 of the regular service includes the provincial governors and the deputy mayor of Seoul; bureau chiefs in the executive ministries are in Grade 2, while section chiefs in the central government are in Grade 3. The administrative assistants and clerks who make up the majority of the regular civil service are assigned to Grade 4 or 5, depending on their qualifications. The special service consists of high-ranking officials appointed without competitive examinations, or on the basis of alternative criteria. This service includes members of the State Council, ambassadors and ministers, deputy ministers of the executive ministries, confidential secretaries of political appointees, judges, and other high officials.

Officers and employees in the regular service are generally appointed on the basis of written examinations and tests of their proficiency. Normally, an appointment is made on the basis of ability demonstrated in a competitive entrance examination, but a special appointment may be authorized after the evaluation of the candidate's abilities, without competitive tests. A ministry or agency receives a list of candidates eligible for an available position and may appoint anyone from the eligible list. Appointments to the special service are made by the president or another official designated by law.

Entrance examinations for the regular service are held on two levels, higher and ordinary. The entrance examinations for Grades 1 through 3 are given by the Ministry of Government Administration; those for Grades 4 and 5 are given by the heads of government ministries and agencies, as authorized by the Ministry of Government Administration. Only a few candidates pass these severe examinations. In 1968, for example, of the 1,909 applicants for Grade 3 positions in administration, finance, and foreign service, only 63 passed. In the same year, only 664 of 12,633 candidates for Grade 4, and 5,400 of 101,076 applicants for Grade 5 positions were successful.[16] In this highly competitive atmosphere, there is no time to waste in the pursuit of learning that will not result in a high examination score. The products of this system are, therefore, sometimes very narrowminded men, capable only of performing well on the entrance examinations.

There has been in recent years a tendency to circumvent the customary entrance examination procedure. Special entrance examinations have been established, purportedly to bring into government service talented men whose qualifications could not be determined in the regular examinations. These special examinations appear in many cases to have been given when a ministry had some reason for preferring a particular candidate. Consequently, those who took the special examinations have fared better than

16. *Hanguk ui Insa Haeng Jong, 1969,* Ministry of Government Administration, pp. 31–32.

the applicants taking the regular examinations; in 1968, for instance, 99 of 187 special applicants for Grade 3 positions were successful.[17] Thus, more men became eligible for top-level civil service positions through the special procedure than through the regular means, and the military regime appears to have created the special examinations solely to bring more ex-soldiers into the civil service at high levels, by way of strengthening its own power.

A study of the Korean political leaders' backgrounds reveals that the cabinet ministers, National Assemblymen, and other high officials of the First Republic (1948–1960) were mostly sons of landholders and former government bureaucrats of the Lee dynasty; they had been engaged chiefly in education and civil service in the Japanese colonial period, and their mean age was 53.4.[18] The typical Korean leader of this period had been born while Korea was under the influence of Japanese expansion and had spent his childhood in an atmosphere of nostalgia for the past, with a family that espoused the conservative values of the Confucian tradition. Having obtained a modern education from the Japanese and achieved a civil service position, he developed a sense of superiority to his parents and friends. On the other hand, he was frustrated by the limited possibilities of social advancement under the highly discriminatory system of colonial rule.

Not all the military leaders who took part in the 1961 coup were of developmentalist orientation; indeed, many of them were similar in outlook to the leaders of the previous regime.[19] The military officers of junior general grade in the late 1950s had had sound training at both Korean and American schools. Many of them had received an extensive formal education before entering the army, and they had numerous opportunities as they rose in rank for contact with intellectuals in civilian life and higher-grade civil servants. While in command positions or senior staff posts in the field, they were gaining valuable skill and experience in both management and planning. Thus they were, relatively speaking, management-oriented. The junior field-grade officers (colonels and majors) of the same period who participated in the military coup had had only limited formal training both before and after they joined the service. Many of them had completed only a secondary education before entering the army, and received six months to a year of basic training at the military academy. In the Korean War, however, they had borne the brunt of combat as platoon and company commanders. After the war, and particularly in the late 1950s, when the opportunities for promotion were exhausted, these officers felt frustrated by a civilian government which they believed was neglecting the moderniza-

17. Ibid., p. 34.
18. Bae-Ho Hahn and Kyu-Taik Kim, "Korean Political Leaders (1952–1962): Their Social Origins and Skills," *Asian Surveys* 3 (July 1963): 305–23.
19. Se-Jin Kim, *The Politics of Military Revolution in Korea,* pp. 77–101.

tion of the military services. At the time of the military revolt, many of them were operations-oriented and apprehensive about their future.

During the rule of the junta, the cabinet and administrative posts were occupied by military leaders of managerial orientation, while operations-oriented men controlled the Central Intelligence Agency; these two factions were eventually reconciled through the establishment of the Supreme Council for National Reconstruction (SCNR) as the supreme organ of the state.

A recent study of the social backgrounds of the political and administrative elite of the Third Republic reveals that they are relatively young and well educated. The median age of a minister is forty-three years; of a deputy minister, about forty-one; of a bureau chief, forty. Ninety percent of them have had at least a university education, and many (including sixty percent of the bureau chiefs) have had opportunities to study in Western institutions, though some stayed there only briefly. The typical member of the elite in the Third Republic was born in the 1920s or 1930s during Japan's occupation of Korea, and received his early education in Japanese schools. But he received his higher education at a Korean college after the Second World War, then perhaps studied in the United States or in Great Britain. Unlike his parents, who suffered psychologically from the Japanese control, he had no particularly strong feelings about the Japanese, whose occupation ended before it could affect him psychologically. After he grew up and became aware of the rest of the world, however, he gradually acquired the spirit and method of nationalism. By the time of his return to Korea, he was firmly imbued with the developmentalist orientation.[20] By and large, the present high-level civil bureaucrats are goal-oriented men, wholly familiar with the techniques of modern management and programming. The military leaders of the Third Republic have discovered that a development program could be formulated and executed more successfully if the goals and managerial orientation of the political elite and the civil bureaucratic elite coincide, and if both have the same basic outlook on the development process.

IV. The Civil Bureaucracy's Other Face

When a serious student of Korean government examines in fine detail the operations of a sufficiently small segment of the civil bureaucracy, he discovers patterns of activity and interaction that cannot be accounted for by its formal structure. Among many others, the following characteristics are peculiar to the Korean civil service.

20. Tong Suh Bark, "Study on Qualifications of Korean Administrative Executives," *Korean Quarterly* 2 (Summer 1969): 19–37.

A. Rank consciousness

An extremely strong sense of hierarchy and rank prevails in the Korean civil bureaucracy, perhaps because of the large number of ex-soldiers in its structure. The wide gap between higher civil servants (Grades 1 through 3) and regular civil servants (Grades 4 and 5) is evident not only in their salaries, but also in such amenities as the size of one's office, its furnishings, and the use of an individual telephone. Officials in Grades 1 and 2 are usually entitled to the use of government-owned, chauffeur-driven automobiles, and usually have more attractive female secretaries than do officials at lower levels. That the bureaucrats' status consciousness sometimes leads to petty behavior is suggested by an anecdote related to the author several years ago; an official in Grade 3 reportedly had the most attractive secretary in his agency, but was compelled by his superior to exchange his secretary for the one employed by the head of the agency, a Grade 1 official.[21] The sense of hierarchy carries over into the work routine as well: subordinates are expected to come earlier and leave later than their superiors, but receive no additional compensation for doing so.

B. Goal orientation

Korean bureaucrats are expected to achieve their officially designated goals by acceptable means. The government may condone an official's irregular acts in office so long as he fulfills his work assignments, but it seldom forgives one who fails to achieve his goal. For example, if the minister of construction is ordered to complete a four-lane highway between Seoul and Taegu, he must do it on schedule, regardless of the means. The same applies to the collection of taxes, construction of bridges, exportation of manufactured goods, or any other government undertaking. Those who fail to meet their goal usually suffer serious losses in promotion opportunities, in their superiors' confidence, and in prestige.

C. Absolute loyalty to the president

In his ideal bureaucracy, Max Weber observed, modern bureaucrats would be hired and promoted on the basis of their skill and efficiency. This system, in which men are loyal to their chosen profession, stands in contrast to the feudal bureaucracy in which men are chosen for their personal loyalty to their lords. Korean civil servants in the Third Republic are expected to be loyal not only to their profession, but also to President Park. The

21. This story was related to the author in Korea in 1968 by a Korean public servant who wishes to remain anonymous.

reason for requiring absolute loyalty to the president is to produce the national unity which, the military leaders contend, is essential for developing Korea from an agricultural into an industrial state. Any dissent, whether constructive or not, is seen as potentially dangerous to the authority of the government.

D. Factionalism[22]

Although the president calls for national unity, there are in the Korean bureaucracy several distinct factions whose members are bound together by personal friendship, school associations, geographical proximity, and professional outlook and experience. One such group supports Jong Pil Kim, the present prime minister and the most powerful figure in the regime next to President Park. Another group, opposed to Kim, supports the policies of Jae Ho Kihl and three other high-ranking officials in the ruling Democratic Republican Party. The anti-Kim faction seemed to have gained some ground when Kim resigned the party chairmanship in the late 1960s, but lost its advantage when Kim re-emerged in politics as the new prime minister. A third group does not side with either faction, but remains a neutral observer of the governmental infighting.

The incident known as the October second affair illustrates the power struggle among the ex-military leaders for control over the civil bureaucracy. After the presidential election of 1971, the president appointed Jong Pil Kim as prime minister and Chi Sung Oh, regarded as Kim's supporter, as minister of home affairs. At the same time, however, Chae-Ho Kihl, one of the leaders of the anti-Kim faction, was named chairman of the DRP policy committee, and his associate Sung Kun Kim became head of the party's central committee. Kihl's power in the National Assembly was further enhanced because of the assistance he had rendered to more than half of the new DRP members of the National Assembly. He had not only helped them in the nominating process, but had also assured their election by his appointment of county officials and police captains at the local level. Thus, he had many friends both inside and outside the party from whom he could expect political favors in return.

Numerous domestic disorders followed these political changes, and there were widespread charges that Chi Sung Oh, who as home minister had control over matters involving local government, had handled these disturbances in a high-handed manner, then shifted local officials about to suit his own liking. In the view of Kihl and his associates, the home minister's reckless actions threatened to destroy not only the power base of the Kihl faction, but that of the DRP as well. The opposition New Democratic Party called for Oh's dismissal as minister of home affairs, alleging that he had

22. *Tong a il Bo,* October 2, 3, and 4, 1971.

exceeded his authority in handling domestic disorders and the replacement of public personnel. The dispute came to a showdown in the National Assembly on October 2, 1971, and in the infighting between the two factions, at least sixteen DRP members revolted and voted with the opposition party, despite President Park's instructions to the contrary. The vote against the government was regarded as a sharp setback for President Park and Prime Minister Kim. Furious, Park ordered the Central Intelligence Agency to investigate and determine who had led the revolt against his instructions. Chae-Ho Kihl and Sung Kun Kim were found to have instigated the rebellion against Home Minister Oh, and were therefore ousted both from the DRP and from the Assembly. Three other members of the anti-Kim faction, including former party spokesman Chang Keun Kim, were suspended from party membership for six months.

V. Conclusion

The Korean civil bureaucracy under the military regime cannot be easily judged as a success or a failure. Some Koreans attribute their country's economic growth to the military leaders' success in establishing an effective and stable government which attracted foreign capital and induced Western-educated scientists and engineers to return to Korea, and in recruiting goal-oriented bureaucrats who could develop and implement well thought-out plans on a national scale. The critics of the military regime, however, argue that Korea made its economic improvement in spite of the military leaders' mismanagement. If there had been no corruption or faulty judgment in the administration, they contend, Korea's gains could have been greater, and they cite countless instances of mismanagement by the leaders of the present regime.

Socioeconomic conditions in Korea have greatly improved since 1961. As the country achieves greater prosperity and national security, its political system will also undergo important changes. The fundamental question about the Korean bureaucracy is not whether it will continue to be efficient and effective, but whether it can ever become a truly responsible public service. Needless to say, the Koreans want their civil bureaucracy to be efficient. But most of them also want it to be democratic—committed to realizing the ideal of popular control over the government, and to serving the needs of the people rather than the ends of a military regime. In short, most Koreans want both efficiency and democracy, but do not wish to choose one at the expense of the other.

PART V

Latin America

THE CAUSALITY OF THE LATIN AMERICAN COUP D'ÉTAT: SOME NUMBERS, SOME SPECULATIONS*

Martin C. Needler

The attempt to understand the causality of the Latin American coup d'état[1] can begin with either of two approaches: the cross-sectional or the longitudinal. That is, it can either examine what factors can be related to differences among the Latin American countries in their propensity to coups; or it can identify the circumstances that seem present when coups occur but not when they don't, in the history of all of the countries, or of groups of them, or even in a single one.

The method should be as quantitative as possible, which may not be very much. Moreover, it should take into account that the parameters it discovers may be characteristic only of certain periods of time, and may not be generally valid. And it should appreciate that even when numerical relationships have been established, different explanatory structures may be created, each equally consistent with the numerical data and with what we believe we know from other sources.[2]

---

* Paper presented at the American Political Science Association meeting, Washington, D.C., September, 1972.
1. The expression "coup d'état" refers here to any extraconstitutional removal of the head of the government, with apologies to purists who would reserve that term only for a particular subcategory of the general type.
2. There is an interesting but unquantified sketch for a model of the coup in John S. Fitch III, "Toward a Model of the Coup d'Etat in Latin America," in Ronald Brunner and Garry Brewer, eds., *Approaches to the Study of Political Development* (New York: Free Press, 1972).

The data base of the present paper consists, in part, of regressions calculated among event-scored and other aggregate cross-sectional data, together with some simple and less formal manipulations of narrower ranges of data; plus the reported results of colleagues' work; plus the general knowledge of the assiduous reader of newspapers. The quantitative work here is necessarily limited in scope.[3]

We take up first the question of the national propensity to coup behavior.

I

It seems clear that, while the possibility of the traditional military coup, or of some other form of removal of a government before the expiration of its legal term, is a salient part of the Latin American political culture taken as a whole, quite different thresholds to military intervention obtain in the different countries of the area. The *golpe* which brought Colonel Banzer to power in Bolivia appears to be the 190th attempted in the 147 years of the country's independence, an average of more than one per year.[4]

The steady economic, social, and political deterioration that has been taking place over the last twenty years in Uruguay, on the other hand, which has—thus far—not provided the occasion for a single military coup, would have prompted at least half a dozen attempts in Bolivia.

The first question that we must examine then is what accounts for the variations among countries in the propensity to stage coups d'état. Let us first take up the hypothesis that a national habit of staging coups can be built up, that is, that a good predictor of the probability of coups is the frequency with which coups have occurred in previous periods.

In fact, quantitative analysis shows this not to be the case. Thus the correlation between the frequency of coups, by country, in the period 1930 to 1965 and the frequency of coups during the nineteenth century (1823–1899) is only .184. But the frequency of coups after 1930 is not even predicted by the frequency of coups in the previous thirty years: the correlation is .025.[5]

In other words, there seem to be changes in a given country from periods of high coup activity to those of low coup activity. This is after all what we know from impressionistic familiarity with the recent history of some of

---

3. One might have used factor analysis and path analysis, and might also have tried lagging longitudinal data. More thoroughgoing work along these lines is being conducted by Douglas Hibbs at M.I.T.
4. See Augusto Montesino Hurtado, "Las Fuerzas Armadas de Bolivia y la Caída de Torres," *Estrategia* (Buenos Aires), no. 12, July–October 1971, pp. 49–51.
5. The data for the numbers of coups are taken from Warren Dean, "Latin American Golpes and Economic Fluctuations, 1823–1966," *Social Science Quarterly*, June 1970. I am indebted to Edward Goff for performing the necessary computations.

Table 1: *Extraconstitutional Assumptions of Power, by Country and Period*

	1823–1965	1823–1899	1900–1929	1930–1965
Argentina	14	8	0	6
Bolivia	26	16	0	10
Brazil	10	4	0	6
Chile	16	9	4	3
Colombia	11	6	3	2
Costa Rica	11	8	2	1
Cuba	7	—*	1	6
Dominican Rep.	28	15	7	6
Ecuador	27	13	4	10
El Salvador	21	13	1	7
Guatemala	21	11	2	8
Haiti	29	13	9	7
Honduras	16	9	5	2
Mexico	21	16	4	1
Nicaragua	16	7	6	3
Panama	5	—*	0	5
Paraguay	21	5	9	7
Peru	27	21	1	5
Uruguay	8	8	0	0
Venezuela	15	9	1	5
TOTALS	350	191	59	100

Source: Warren Dean, "Latin American Golpes and Economic Fluctuations, 1823–1966," *Social Science Quarterly*, June 1970.

* Not independent during the nineteenth century.

the countries of the area. Traditionally turbulent Mexico has seen no successful extra-constitutional seizures of power since 1920, and the constitutional succession has been unbroken in previously undemocratic Venezuela since 1958. Even Bolivia, the all-time Latin American champion in the frequency of attempted coups,[6] enjoyed a twelve-year period, from 1952 to 1964, under the rule of the Movimiento Nacional Revolucionario in which presidents served out their terms and were followed by elected successors.

The change may take place in the other direction, too, as the history of Argentina shows. The Argentina whose politics were peaceful and orderly in the early twentieth century began in 1930 a new period of military intervention in politics in which military governments and the seizure of power from elected civilians have become as commonplace as in the most stereotyped Central American countries.

Similar mutations may still take place. Even in the countries which seem

6. Although yielding to Haiti in numbers of successful coups.

to have placed the era of the coup d'état behind them, such as Mexico and Chile, or Uruguay and Costa Rica, military intervention remains even today a possibility. Thus in Mexico, for example, the attempt was made in May of 1971 by conservative forces within the regime to force President Echeverria out of office, apparently using tactics of noncooperation with the president similar to those employed to force Ortiz Rubio's resignation forty years before. The movement failed after a meeting of military commanders decided to back the president.[7]

Likewise, it is no secret that the Chilean armed forces stand ready to execute a coup against Salvador Allende if the president should take actions seriously violating constitutional norms, and the question was discussed within the armed forces prior to Allende's election to the presidency.[8] In Costa Rica right-wing opponents of President Figueres tried to organize the removal of the president in reaction to his plan to establish diplomatic relations with the Soviet Union, and at the beginning of 1972 the chief of the CIA in Costa Rica was recalled, rumor has it for complicity in this plan. In May 1972, the Uruguayan armed forces were described as ready to remove President Juan Bordaberry if he was not able to secure congressional support for the continued suspension of constitutional guarantees, which he was fortunately able to do.

In the short term, that is, the national propensity to coups may be a constant; but it is not a constant over the whole period of national history since independence.

II

Before continuing with the examination of the varying propensity to coups of different countries, let us look at what we know about the occasions for coups d'état, or at least the characteristics of the periods at which they are likely to occur. On this point, as this writer found in a previous study, coups occur particularly in an election year and/or when economic conditions are deteriorating.[9] The point that coups tend to occur in election years seems to have been taken so much for granted that apparently no one has taken the trouble to study it further. Perhaps, as one scholar wrote, "Everyone knows that most *coups* occur at such times." [10]

Several recent studies seem to have confirmed amply the thesis that revolts

7. See M. C. Needler, "A Critical Time for Mexico," *Current History*, February 1972.
8. See Miles Wolpin, "Chile's Left: Structural Factors Inhibiting an Electoral Victory in 1970," *Journal of Developing Areas*, January 1969.
9. M. C. Needler, *Political Development in Latin America* (New York: Random House, 1968, pp. 61–62).
10. David C. Rapoport, "The Political Dimensions of Military Usurpation," *Political Science Quarterly*, December 1968.

are more likely to occur when economic conditions are deteriorating. Kristin Parrott found that thirty-two of the sixty-eight coups she studied occurred during "an economic recession or crisis." [11] Warren Dean found that the incidence of coups d'état during the nineteenth century corresponded closely with the weakening of economic conditions as measured by the decline in the volume of foreign trade.[12] Gilbert Merkx has found a correlation between armed rebellions in Argentina and the worsening of economic conditions.[13] According to Frank Bonilla, the major attempts to overthrow Venezuelan governments have followed economic downturns.[14] And Egil Fossum found that from 1907 to 1966 proportionately twice as many coups occurred when the economy was deteriorating as when it was improving.[15]

The occurrence of coups d'état during periods of economic constriction may call for a little more elaboration than seems necessary on the surface. We may be, and probably are, dealing with several different types of causal connection which, though interrelated, have rather different implications for theory. The direct connection between economic hard times and the removal of a government may, for example, lie in what happens to the military budget. Víctor Villanueva found that the Peruvian governments which had finished their terms from 1912 to 1970 were those which had not reduced the military share of the government budget; the proportion of the budget going to the armed forces had been reduced by all of the governments that were overthrown.[16]

Alternatively, economic difficulties may be regarded as a demonstration of the government's incompetence, and it is removed as a kind of military "vote of no confidence." The typical reason for the overthrow of Velasco Ibarra in Ecuador (which has occurred four times)—although not for his most recent removal from office, which took place in 1972—has been his economic bungling.

Probably more relevant is the third type of connection between an unwelcome state of the economy and the premature retirement of the president: that social conflict increases during a time of economic downturn. For example, it seems clear that the aspect of the economic situation which helped precipitate the removal of General Ongania from the Argentine

11. Kristin Parrott, "The Latin American Coup d'Etat," unpublished paper, University of New Mexico, May 1972.
12. Warren Dean, "Latin American Golpes and Economic Fluctuations, 1823–1966," *Social Science Quarterly,* June 1970.
13. Gilbert W. Merkx, "Economic Cycles and Armed Rebellions in Argentina, 1879–1956," mimeographed, University of New Mexico, October 1968.
14. Frank Bonilla, *The Failure of Elites* (Cambridge, Massachusetts: Massachusetts Institute of Technology Press, 1970), p. 38.
15. Egil Fossum, "Factors Influencing the Occurrence of Military Coups d'Etat In Latin America," *Journal of Peace Research,* Fall 1967, p. 237.
16. Víctor Villanueva *¿Nueva Mentalidad Militar en el Perú?* (Lima: Editorial Juan Mejía Baca, 1969), p. 194.

presidency in 1970 was that the necessity, as he saw it, of holding down wages during a period of inflation led to public disturbances. These brought the military, used in the role of gendarmes that they so abhor into direct conflict with urban workers.

The heightened disorder and social conflict of an election period may to some extent be responsible for the overthrow of governments at such a time, although it seems clear that most coups which occur in election years are designed to head off the victory of a candidate unacceptable to the military. (Or at least unacceptable to some significant element of it. Other sectors can usually be induced to follow because they are reluctant to divide "the institution.") A wide variety of objections to the probable, or actual, election winner may enter into the decision to seize power, many bearing directly or indirectly on the self-interest of the military itself.

What appears from all this is that in Latin America the military seizure of power is an immediate alternative to elections or the functional equivalent of motions of impeachment and votes of no confidence; but then all we are saying by concluding this is that coups occur when the actual performance of a present government, or the probable performance of a future government, is thought unsatisfactory.

The critical question, however, must accordingly become not why governments should not be completely satisfactory, since after all perhaps most governments disappoint, but why are not disappointing governments always removed in this direct fashion; in other words, why do coups often *not* take place? What factors inhibit the occurrence of coups?

III

Now a satisfactorily high inverse correlation with the frequency of coups d'état in the different countries in Latin America can be found to exist with a number of variables; this was pointed out some years ago by Robert Putnam.[17] However, these are all variables which measure, under one guise or another, the factor "social and economic development" and thus appear to be, in effect, a variant form of our old friend and unshakable companion, gross national product per capita, which usually itself provides the greatest correlation with numbers of coups of any of these "development" indicators. As Karl Deutsch pointed out long ago,[18] there is really no point in considering separately such indicators as literacy, newspaper circulation, ownership

17. Robert Putnam, "Towards Explaining Military Intervention in Latin American Politics," *World Politics,* October 1967. See also José Nun, *Latin America: the Hegemonic Crisis and the Military Coup* (Berkeley: Institute of International Affairs, 1969).
18. Karl Deutsch, "Social Mobilization and Political Development," *American Political Science Review,* September 1961.

of radio receivers, urbanization, and so on, since they all load on the same factor of economic development and are just as well represented by GNP per capita figures.

Of course the problem remains that of deciding how and why development serves to inhibit the frequency with which coups occur. Clearly, there are a variety of possible causal relations:

1. The higher level of economic well-being represented by a higher GNP per capita figure leads to greater satisfaction with the status quo and unwillingness to contemplate change;

2. Higher economic level means more government capacity to meet demands for services and to deal with social problems effectively before they reach an acute state;

3. Either of the foregoing relationships may, as Seymour Martin Lipset has argued,[19] create, maintain, and augment feelings that the political system is legitimate since it appears to be operating successfully;

4. The rise in social mobilization leads to heightened political awareness and sophistication and the transcending of the state of political culture in which violent methods of settling political problems are regarded as acceptable. This is a major point made by S. E. Finer.[20]

All of these factors may be operating, or none of them, or it may be that yet other relationships are more important. However, the ones listed are a priori most plausible and are listed to remind the reader that the discovery of a relationship between two variables, after indicators of them have been quantified, does not in itself constitute a causal explanation.

The clue to the direction in which explanation lies appears to be provided by another variable which, while associated to some extent with the GNP-per-capita complex, is correlated more highly with the frequencies of coups than any of the development indicators, including GNP per capita itself: that is the percentage of the population which is Indian. This factor might correlate highly ($r = .615$) with the frequency of coups in the nineteenth century, however, because it serves as a "stand-in" for GNP per capita figures which we do not have for that period. However, it also correlates much more highly with frequency of coups for the recent period, 1930–1965 ($r = .489$) than does the GNP per capita figure ($r = .362$), although the correlation disappears, and in fact becomes negative, for the period 1900–1929 ($r = -.254$). I hesitate to take these correlations per-

19. Seymour Martin Lipset, *Political Man: The Social Bases of Politics* (Garden City, N.Y.: Doubleday Anchor edition, 1963, chap. 3).
20. S. E. Finer, *The Man on Horseback* (New York: Praeger, 1963), chaps. 7–9.

fectly literally, since the figures for the percentage breakdowns of population by racial category are so unreliable,[21] among other reasons, but let us
assume that there is something to this relationship and investigate its implications further, looking first at the relation between coup frequency and
percentage of Indians, and then at the special character of the period
1900–1929, in which relationships visible before and following do not seem
to hold.

IV

For those not familiar with Latin America the usual caveats have to be
entered before one discusses questions of "race." That is, as it is used in
Latin America, "race" has only in part a physical or genetic denotation,
referring primarily to culture and way of life. It should also be noted that
the dimension of Indian/European is not the only axis of racial differentiation
to be found in the area, but is in several countries replaced or supplemented
by the dimension black/white.

Considering for the moment only the Indian/European dimension, we
must somehow account for the fact that, as table 2 shows, the frequency
of coups has been greater in the heavily Indian countries.

Table 2: *Incidence of Coups in Strongly Indian Countries,*
1823–1965 and 1930–1965

	Percent of Indian Population	1823–1965	1930–1965
Bolivia	63	26	10
Guatemala	53.6	21	8
Peru	45.9	27	5
Ecuador	39	27	10
Mexico	30	21	1
El Salvador	20	21	7
Latin American average		17.5	5

Sources: (1) For population, Yvan Labelle and Adriana Estrada, *Latin America in
Maps, Charts, Tables* (Cuernavaca: Center for Intercultural Formation, 1963), p. 77.
(2) For coups, Warren Dean, "Latin American Golpes."

One possible explanation for all this is that a high degree of social
mobilization in a society acts to inhibit the extraconstitutional seizure of
power. Given the suppressed condition of the Indian masses, the persistence

21. I have used the figures given in Yvan Labelle and Adriana Estrada, *Latin
America in Maps, Charts, Tables* (Cuernavaca: Center for Intercultural Formation, 1963), p. 77.

of Indian languages, and the existence of a major communications gulf between the Indian and non-Indian sectors of the society, the fact that a high proportion of the population can be considered Indian may well be a better indicator of a low degree of social mobilization than more frequently used measures as literacy or GNP per capita. That is, it is plausible to argue that although the characteristics of Indian populations do not cause coups d'état to occur, their lack of social mobilization and integration into national society may *fail to inhibit* the occurrence of coups.

Although this may be part of the explanation, it should be noted that the correlation between a social mobilization indicator such as GNP per capita and the Indian percentage, while certainly there, is not very high (it was .26 in my data). In his factor analysis of a large number of aggregate data indicators, Philippe Schmitter found that indicators tapping the dimension of the Indian character of the society loaded on a different factor from his modernization/development indicators.[22] He also found, by the way, that the Amerindian factor, and not the modernization/development factor, correlated with the incidence of violence (.35), although his violence index was derived from instances of riots, civil wars, and so on, and thus did not necessarily represent the same dimension as frequency of coups d'état, the problem that we are dealing with here.

It might perhaps also be pointed out by way of explanation that, because of the pre-Columbian development of the major centers of Indian civilization in the high mountain areas, the countries with the largest Indian populations are those with difficult terrain that were most cut off, until the coming of jet traffic, from the outside cultural and political influences that promoted the spread of democratic ideas; and that these were also brought by the European immigrants of the nineteenth and early twentieth centuries, who for the most part did not come to the Indian countries.

To my mind, however, the most convincing structure of explanation linking the frequency of coups and the Indian character of a society lies not in these factors, but in the attitudes and behavior which the existence of a subject caste of Indians generated in the non-Indian sectors of society. It is clearly in societies of this character that the gap between political norm and social and political reality, so marked in Latin America, is at its greatest. It is clearly in societies of this type that it is least possible to give credence to the premises of social equality, political participation, and majority rule that are implicit in the norms of the formal constitution. It is in these societies, that is, that democratic institutions are least legitimate. The fact of subordination of the Indian masses, at least a substantial minority if not a majority, can clearly not be acknowledged openly in constitutions almost always based on the prevailing Western ideals of representative democracy.

22. Philippe Schmitter, "The Ecology of Political Change in Latin America: A Cross-National Exploration," mimeographed, March 1970.

The message may be there, for those who know how to read between the lines. In the typical Ecuadorean constitution (there have been a great many), for example, citizens are those born in Ecuador *who are able to read and write*. This type of device is common in other contexts: there are minimum height standards for the officer corps of the Peruvian army that just happen to exclude perhaps ninety percent of the shorter indigenous population.[23]

To a citizen of the United States this sort of thing is reminiscent of the voter registration tests of recent memory in the South, the purpose being the same: to bring a legal structure based on premises of equality, majority rule, and constitutional practice into operational compatibility with a racist society. It is certainly not coincidental that until twenty years ago, at least, the South was the most violent region of the United States, as the Indian societies are the most violent in Latin America. On the one hand, the formal institutions of democracy are taken least seriously, because everyone knows that, past a certain point, they are fraudulent. And on the other hand force must be kept in constant readiness, and occasionally used, against the possibility of an insurrection by members of the lower caste.

The case of Haiti is rather similar to that of the Indian societies for those periods of its history when the mulatto elite has been in control. In the periods of political participation by the black masses, among whom the democratic constitutional norms of the Western cultural complex have not penetrated, on the other hand, kings and emperors and superhuman life-time presidents are as legitimate—no, more legitimate—than a constitutional ruler would be.

V

As was noted above, the relative frequency of coups by country differs from one period of time to another, and during the early part of the twentieth century the Latin American countries, and especially those with large Indian populations, were more stable and less coup-prone than they were during the second third of the present century or during the nineteenth century. There are different ways in which this phenomenon can be accounted for, each of them probably containing some truth.

A traditional political analysis would point out that there was economic prosperity around the turn of the century—GNP per capita, if we had the figures, would doubtless show a rise—and this made possible all the good things that come with higher GNP per capita, political stability among them. If the stability of the Indian countries was especially marked during this period, this might be attributed, from this perspective, to the fact that

23. Villanueva, *¿Nueva Mentalidad Militar en el Perú?*, p. 254.

several of these mountainous countries are also the ones that exported the metals and other minerals whose prices were particularly high at the time.

A Marxist interpretation might be that the early twentieth century was a period of the consolidation of the rule of a single dominant class (the externally dependent bourgeoisie) in contrast both with the previous century, in which there had been conflict between feudal and rising bourgeois elements, and with the middle of the twentieth century, which has seen conflict between the dependent sector of the bourgeoisie and the national bourgeoisie, and conflict between the bourgeoisie as a whole and a rising working class.

From the vantage point of the stress on ideologies and values which produced the discussion of legitimacy above, it could be argued that because of the racism which dominated Western sociology at the turn of the century and was incorporated into dominant ideologies in Latin America at the time as part of the positivist synthesis, for the first time the actual social structure in the Indian countries could be openly defended in terms that were intellectually respectable; the tension between norm and reality could be reduced; and the existing political order could be believed to be legitimate.

VI

Professor Stepan's conclusion from his Braziliar. study, that governments are less likely to be overthrown the larger their electoral margins,[24] does not emerge clearly from the data on all of Latin America that I was able to analyze. Because of the very large number of governments involved and the limited time available to secure complete information, I have carried the investigation back only a little over ten years, to January 1, 1962.

Table 3 shows how governments of different types left office in the ten-year (actually, ten and one-half year) period 1962–1972. Governments elected with a large plurality did have a slightly lower chance of being overthrown than those elected with a small plurality (that is, one of less than four percent of the vote cast) but the difference was small and may not be significant. The table does suggest, however, that the election of the head of government may well increase the legitimacy of his position, and thus his chance of completing his term. Whether or not the difference in ability to remain in office between governments elected with large and those elected with small majorities is regarded as significant, there does seem to be a smaller likelihood of the overthrow of elected governments than of those not elected.

24. Alfred Stepan, *The Military in Politics: Changing Patterns in Brazil* (Princeton: Princeton University Press, 1971), pp. 80, 250. This is an excellent study, especially creative and insightful in its discussion of the legitimacy question.

Table 3: *Governments Leaving Office, January 1, 1962–June 1, 1972*

Type	Total No.	Finished Term	Overthrown	% Overthrown
A. Elected w/large majority	26	19	7	26
B. Elected w/small majority (margin of less than 4%)	10	7	3	30
C. Vice-presidents who succeeded	6	3	3	50
D. Irregular origins	16	10	6	37
Totals	58	39	19	34

Vice-presidents attempting to serve out the term of a president who died or was removed from office, for example, had only an even chance of completing the term. In fact the odds would be even less favorable if the case of Pedro Aleixo, left out of table 3, were included. It might be expected that a vice-president would not be allowed to succeed a president who is removed by the military, but in Brazil Pedro Aleixo was not allowed to succeed even when President Costa e Silva was incapacitated by a stroke, notwithstanding the fact that Aleixo had been placed in office by the military themselves under a constitutional arrangement they had just themselves decreed. And in two of the three cases in which vice-presidents completed the term, those of Guido in Argentina and Guerrero in Nicaragua, the new presidents served only as front men for the military commanders.

The hypothesis of electoral legitimacy seems also to bear on the survival of the "irregular" governments, those not themselves popularly elected. The ten governments in this category which handed over office peacefully to elected successors all regarded themselves explicitly as provisional governments whose function it was to prepare for elections. The saying "only the provisional lasts," applied by the French to the institutions of the Third Republic, applies with particular aptness, though in a rather different sense, to these governments. Of the six governments in this category that were overthrown, four were military regimes which planned on staying on without holding elections (Ovando, Torres, Onganía, and Levingston). The fifth government, that of Reid Cabral in the Dominican Republic, was a civilian regime that was apparently planning controlled elections that would perpetuate itself in power. The sixth government, the Ecuadorean military junta of 1963–65, was unable to decide if it should stay in office without elections or hold elections, and if the latter, whether there should be an official candidate, and if so, who he should be. However, it should be acknowledged that the long-lived regimes of Haiti and Cuba have found sources of legitimation other than elections, while the elections which constitute the homage of hypocrisy rendered by some of the hemisphere's military rulers to its democratic ideals have been, most notably in the case of the Brazilian military, the most transparent of formalities.

VII

It appears probable, then, that a government enjoys greater legitimacy if it was popularly elected, perhaps even more if it was elected by a large margin. It thus partakes of what might be called "democratic legitimacy." At the same time, it should be emphasized, in line with the discussion of the Indian countries above, that the capability of democratic norms for legitimizing any particular government is lower in those countries where democratic norms themselves lack legitimacy because they are so clearly out of consonance with the premises on which society is actually constructed.

It is of critical importance that in Latin America two alternative sources of legitimation of regimes exist to which it is possible to appeal against the democratic legitimacy of elected governments. The first of these is Christianity, which, for many Latin Americans, asserts claims that must be given precedence over those of democratic legitimacy in some cases. Thus the litany of the manifestoes issued by de facto governments seizing power extraconstitutionally draws heavily on rhetoric about Western Christian values, generally understood today politically as implying anti-Communism. On occasion this is not simply post facto rationalization, but may actually play a part in the insurrectionary movement. In the overthrow of Juan Bosch, for example, the role of his government in introducing a new constitution providing for separation of church and state provided a genuine ground of opposition for many devout Dominicans.

The other principal source of non-democratic legitimacy is "revolutionary legitimacy." [25] The origins of the national identity of the Latin American nations in successful revolutions is universally taken to mean that a successful revolution creates its own legitimacy. There have indeed been judicial decisions to that effect, such as the one handed down by the Havana judge who heard the young Fidel Castro's plea that acts of the Batista government be declared illegal since the government was itself illegal. The appeal to revolutionary legitimacy constitutes at least the formal basis of the claims to obedience of de facto governments. A merely formal claim of this character does not however itself create the sort of legitimacy that

---

25. On this point, see Samuel P. Huntington, "The Role of Military Influence in Foreign and Domestic Policy: Europe, Latin America and the United States," in *The Working Papers from the 1970 Atlantic Conference* (New York: Center for Inter-American Relations, 1971), pp. 76–79.

inhibits further attempts at coups d'état. A deeper revolutionary legitimacy in this sense may be acquired if the regime embarks on what most people would regard as an authentically revolutionary program including, for example, land reform and the nationalization of private property. In such cases the government derives support not only from popular enthusiasm for the measures themselves but also from the feeling that as a genuinely revolutionary government it has earned the right not to be bound by preexisting norms, such as, for example, the obligation to hold elections. This is clearly the case with the present Cuban government, and may also be true of the current regime in Peru, although it is rather too soon to tell. In Mexico, and in Bolivia between 1952 and 1964, the revolutionary regimes did hold elections although it seems clear, at least from our present vantage point, that the regimes were solidly enough entrenched and enjoyed enough revolutionary legitimacy that they could have dispensed with elections had they so chosen. But in both cases a strong component of the programs of the revolutionary movements was precisely objection to the fact that previous governments had not held fair elections, so the holding of such elections became an integral part of the revolutionary creed.

It may well also be that a further source of the legitimation of regimes is, as Lipset has argued, that approval of a specific government's performance, for example as regards the economy, may, if sustained over time, be transmuted into feelings of legitimacy toward the political system within which successive governments serve. If this is the case, it should be especially true, according to what we know about learning theory, where the government in question is the first to take office under a new political system. The successful performance of a succession of governments in a new political system may thus serve to attach legitimacy to that system by processes of "imprinting" and "reinforcement." This approach provides one of the several alternative possible explanations of the peculiar stability of the Chilean political system. The new Chilean state enjoyed "a precocious economic growth immediately after the proclamation of independence, in contrast with the long depression which was the fate of the young Latin American republics after the wars of independence," [26] and was subsequently successful in wars with its northern neighbors.

VIII

To recapitulate the principal points made in the rather rambling discussion above:

1.   The propensity to stage coups d'état differs among the Latin American countries. However, these differing national propensities have not re-

26. Regis Debray, *The Chilean Revolution: Conversations with Allende* (New York: Random House Vintage Books, 1971), p. 26.

mained constant throughout national history and will presumably continue to change.

2. Coups occur especially around election times and/or in periods of economic deterioration. This need mean no more than that they represent adverse judgments on the performance of the present or prospective government. In itself, this does not help to explain why coups occur and it seems more profitable to look for factors that inhibit the occurrence of coups.

3. Such a factor is socioeconomic development. More significantly, coups are more frequent where there are large Indian populations.

4. Various explanations can plausibly be adduced for these findings. Several considerations induce this author to stress those that refer to cultural development and political attitudes, especially those relative to democratic legitimacy.

REVOLUTION FROM WITHIN?
MILITARY RULE IN PERU SINCE 1968

*Luigi R. Einaudi**

Introduction

In 1871, Karl Marx wrote to Ernst Kugelmann that the fate of the Paris Commune demonstrated that henceforth popular revolutions could not succeed without first smashing the increasingly powerful repressive arm of the modern bureaucratic state: the armed forces.

In 1954, the dominant elements of the Guatemalan military, acting with the support of the then foreign policy of the United States, helped to overthrow the elected government of Colonel Jacobo Arbenz and reversed what was beginning to be an increasingly radical agrarian reform program.

In 1959, a triumphant Fidel Castro took what he considered the first key steps toward consolidating his power by declaring himself Commander-in-Chief of the Cuban armed forces and placing his brother and his most trusted guerrilla *comandantes* in charge of a program of restructuring the Cuban officer corps, thereby ensuring himself against internal counter-revolution. Only then did he turn to the agrarian reform that was to prove the symbol of the Cuban revolution.

In 1968, in direct contrast to these previous events and to the generalizations commonly drawn from them, Peru's armed forces, acting (in exemplary bureaucratic fashion) under the command of the military chief of staff and

* This paper was originally delivered at the 1971 Annual Meeting of the American Political Science Association, Conrad Hilton Hotel, Chicago, Illinois, September 7–11, and is reprinted with permission from *Studies in Comparative International Development*, Vol. VII (Spring) 1973, #1, 71–87 with a new postscript.

the commanders of the three services, seized power from an elected but ineffective liberal democratic regime, nationalized without compensation the local subsidiary of Standard Oil of New Jersey, and set under way a revolutionary process that has included Latin America's most radical agrarian reform since Cuba.

What makes these Peruvian events doubly puzzling is that the revolution was led by many of the same officers who in 1965 had destroyed in blood the guerrillas of the Movement of the Revolutionary Left (MIR) led by Luis de la Puente. Could it really be, as Fidel Castro colorfully put it last August with his characteristic sense for the jocular, that "the fire has broken out in the firehouse"?

This essay seeks to address Castro's question about Peru by looking first at the firehouse and then at the fire. First, it considers briefly the military forces of Peru and their political behavior, which have so strikingly contradicted the stereotypical view of the military as committed to the preservation of the status quo through repression. It then considers even more briefly some of the innovations the military government has attempted to bring to Peru since 1968 in order to conclude by suggesting some implications of these events for strategies of change in general.

The Military Sources of the 1968 Revolution in Peru[1]

The evolution of the political style of the Peruvian military is complex and halting, as befits the behavior of an organized bureaucracy in a rapidly, if unevenly, developing society. It could nonetheless be argued that events since October 3, 1968, mark a qualitative shift in military participation in politics from a generally cautious political stance dedicated primarily to arbitrating between the policies and leaders advanced by civilian political groups to an even more dominant role, placing military leaders directly in policy-making positions to the point of virtually excluding civilians. Military rule, previously generally conservative and caretaker in style, now claims to be revolutionary, and is introducing innovations in public policy that, when considered in the past, were as regularly rejected.

The origins of this pronounced leftward shift in the style and substance of military political activity are rooted in a complicated combination of institutional military factors, of personal experiences of members of the military officer corps, particularly in the army, and of the course of Peruvian society in recent decades. What happened may be summarized at one level by

1. Military politics in Peru has been one of the author's primary intellectual concerns for some years, leading most recently to the Peruvian section of *Latin American Institutional Development: Changing Military Perspectives in Peru and Brazil* (R-586, The Rand Corporation, 1971), written with Alfred C. Stepan III, who wrote on Brazil. The discussion that follows is largely drawn from the document, particularly pp. 16–31.

saying that military leaders began to perceive national security problems as extending beyond conventional military operations. They did so in large part because many of the existing social and economic structures seemed so inefficient or unjust as to create the conditions for, and give legitimacy to, revolutionary protest and hence constitute a security threat. Even conservative officers came to feel that these conditions could ultimately become a threat to the military itself as an institution.

In the late 1950s and early 1960s, Peruvian officers increasingly saw their society as caught up in a fundamental, long-term crisis that threatened them both as military men and personally as members of an often hard-pressed middle class. Land invasions, guerrilla movements, and acts of political terrorism were seen as the top of an iceberg of inexorably mounting social pressures caused by exploding populations that would, in the long run, overwhelm traditional social structures. Military officers in the national war college, *Centro de Altos Estudios Militares* (CAEM), increasingly studied a wide range of social problems. These included questions of land reform, tax structure, foreign policy, and insurgency, and involved the formulation of policies and reforms the military felt necessary to ensure stability. The result was that military policy became much more closely linked to political policy than it had been in the past.

The guerrilla experience of 1965, though successfully controlled in military terms, underscored to the military the importance of social change. It also raised fundamental doubts about the capacity of civilian-directed efforts to achieve that change—despite the fact that the military junta of 1962–1963 had helped install Fernando Belaúnde as president with the hope that he would prove to be a successful reformer. By 1966, military men were ready to judge Belaúnde a failure, the more so as many of them envied the power and the activity of the civilian professionals around Belaúnde, many of whom were financially rewarded beyond the highest expectations of general officers although often far less competent.

The impotence of the Belaúnde government, together with the continuing presence of the aging and by then largely complacent Apristas, heightened the military's anger. Their frustration was continuously fed by incidents of social rejection by the pretentious "whiter" social elites of the coastal cities, and by the antimilitary arrogance of the United States, whose efforts to promote the Alliance for Progress often appeared to many army officers as anti-Peruvian meddling limited to the defense of U.S. economic interests.

These shifts are of sufficient interest to warrant their more detailed examination, in both theory and practice, from the viewpoint of the Peruvian military leadership, as traditional military resentment of civilians, plutocrats, and foreigners became focused increasingly on organized political parties, which were in turn viewed as hopelessly committed to an unjust social, political, and economic order. These views, together with the wider concerns over economic backwardness and social instability, gradually permeated the training and operations of two major institutions: the military schools, and the military intelligence services.

Theoretical Perception of Threats:
The Military School System and the CAEM

Ever since the founding of the Center for Military Instruction (CIMP) and the Center for Higher Military Studies (CAEM), in the first years after World War II, military education had improved and expanded to include socioeconomic concerns.[2] The CIMP opened in 1948; it pulled together military education under a single command, laying the basis for greatly expanded emphasis on continuing post-academy training. The CAEM, which opened in 1950, offered a one-year course largely devoted to social, economic, and political problems to selected classes of colonels and generals. Moreover, officers were encouraged to follow specialized military or civilian courses both in Peru and abroad, usually at government expense. Members of the officer corps had studied military affairs in Europe and the United States since the turn of the century, often winning recognition as the best foreign students. In the 1950s and 1960s, Peruvian officers studied economics under United Nations auspices with ECLA in Chile, attempted unsuccessfully to import the Belgian Catholic sociologist Frère L. J. Lebret to teach, and made innovative policy suggestions to the conservative Prado government, some of whose members in turn became concerned at "Communist" infiltration of the CAEM.

As William F. Whyte has suggested, most Peruvians do not believe that success in life is based on merit.[3] In the military, however, the emphasis on professional training and education in the promotion process, particularly since World War II, has made its members perhaps the most merit-oriented within the state bureaucracy, if not the entire society. All navy and air force officers and more than ninety percent of all army officers are Academy graduates. The continuing value of education in the Peruvian military career may be inferred from the fact that, of the division generals on active duty between 1940 and 1965, no fewer than eighty percent had graduated in the top quarter of their class at the Military Academy. In addition, the expansion and improvement of advanced military training after 1945 introduced a new element of competition into the officer corps and improved the life chances of officers previously stymied by the promotion system's dependence on class

2. A readily available (but sketchy) summary of the structure and function of military educational institutions, emphasizing the role of General José del Carmen Marín in the founding of CAEM, will be found in Luis Valdez Pallete, "Antecedentes de la nueva orientación de las Fuerzas Armadas en el Perú," *Aportes,* no. 19, January 1971.
3. William F. Whyte and Graciela Flores, *Los Valores y el Crecimiento Económico en el Perú* (Lima: Senati, 1963).

standing at the time of graduation from the Military Academy. The final academy class standings, determined on the basis of combined academic and discipline performances, were often stifling to bright men who were deficient in discipline or conduct, but who were in the 1950s and 1960s given new chances to prove themselves in advanced education, much as the development of the air force had in the 1920s and '30s provided an outlet for energetic and talented officers who had been "burned" in the regular army.

The CAEM is probably the most important center for the development of Peruvian national security strategy.[4] Many of the changed military perceptions have crystallized in CAEM studies and class exercises. Of the first nineteen cabinet ministers after the 1968 revolution, thirteen were CAEM graduates, including the prime minister and the chief of the Council of Presidential Advisors (COAP). The director of the CAEM is a general appointed by the Joint Command (*Comando Conjunto*) of the Armed Forces. The director has a small staff and three departments and three directorates under him. The departments include the *Office of the Deputy Director,* who is in charge of administration and through whom the three directorates report, the *Academic Council,* made up of the heads of the directorates and selected professors, and, finally, an optional body called the *Consultative Council,* which may be convened at the director's discretion to study special problems. Of the directorates, the *Academic Directorate* is responsible for plans, academic programs, and the actual content of instruction. To it are assigned the students, mainly officers with the rank of full colonel, known as participants. The *Directorate of National Strategy and Special Studies* studies contemporary problems of national security, special problems, and the strategies of foreign powers, including the United States. The third directorate, *Research and Development,* is concerned exclusively with the future.

CAEM courses in 1970 opened with an introductory study of methodology, sociology, and similar general principles.[5] This was followed by the first major curriculum segment, the study of national reality. This segment of instruction was followed by the analysis of national potential, defined as constantly moving and changing. The contrast between reality and potential establishes national objectives which are to eliminate the differential between reality and

4. National planning, conceived less as general theory and more as related to immediate government policies, is carried out elsewhere, of course, primarily at the National Planning Institute (INP), originally established with mixed civilian and military personnel by the 1962 military junta. Civilian institutions, including universities and the research-oriented Institute of Peruvian Studies (IEP), are important primarily through the impact on CAEM doctrine achieved through individual faculty members.
5. This paragraph, like the preceding one, is partially based on an interview with General Augusto Freyre García, director of National Strategy and Special Studies at the CAEM.

potential. National problems are of course studied from economic, social, military, and psychological viewpoints, each of which was represented in the second major part of the course, the study of national strategy. National strategy consists of the actual programs designed to attain national objectives. The final portion of the course is devoted to individual case studies. These studies, drawn up by the participants in the CAEM, benefit from the experiences not only of the military participants but of the civilian as well. Although occasional civilian students attended CAEM classes as early as 1961 and 1962, it was not until the mid- and late sixties that the numbers became significant on a routine basis. In 1971, sixteen students out of forty-three were civilians.

The CAEM has consistently taught, since its founding in 1950, that, in accordance with Article 213 of the Peruvian Constitution, the military must defend national sovereignty. Specifically, this is defined as an obligation to increase Peru's capacity for maneuver vis-à-vis the outside world, and particularly the United States. Recognition of the Soviet Union, coupled with some trade, is in harmony with this interpretation of the constitutional mandate. Similarly, the constitutional prescription for the maintenance of order is now interpreted at the CAEM as the need to ensure an order conducive to "national well-being," that is to say, the well-being of all Peruvians, not just of the dominant social classes.

To argue in favor of economic development and social justice is not, however, to have a clear plan for how to bring them about. Military men have a powerful impulse to see politics in a fundamentally apolitical light. This contradiction is exemplified in the opening paragraph of the action program adopted in 1944 by a secret military lodge, the Comando Revolucionario de Oficiales del Ejército (CROE):

CROE has no political implications of any kind. It is a revolutionary organization of the officers of the Army who aspire to lead the country within a democratic and strictly constitutional order.[6]

The failure to realize that revolution involves politics, though a generation old, is still typical, as is CROE's stress on the need for "morality" in public life. As recently as 1970, General Velasco was proclaiming that he was a "soldier and a revolutionary, not a politician." [7]

To say that the traditional military prescription for good government is morality, discipline, and patriotism is to be reasonably close to fundamental

6. Victor Villaneuva gives the complete text in the Appendix to the 1956 Peruvian edition of his classic *La Tragedia de un Pueblo y un Partido.*
7. As the President put it in a speech in Trujillo, October 11, 1969: "We are not politicians, professional or otherwise. We are soldiers and we are revolutionaries." (Version issued by the Oficina Nacional de Información, p. 36.)

old-line military attitudes. There has also always been some tension, however, between these views of politics and the military's sense of inferiority in cultural and social matters. "When a general met an ambassador, he turned red in the face and trembled," said a former minister of war, one of Peru's leading military intellectuals.[8] Traditionally, the thought persisted that successful politics might require more than could be brought to it by the military.

Peruvian officers' attitudes toward civilians and toward politics have historically combined into a powerful dislike for civilian politicians. Their attitude toward politics, taken with the self-image of discipline and efficiency, leads some officers to believe that they are the elect and must lead the nation. Typically, however, these same attitudes also lead to another and somewhat contradictory sense of contempt for officers who "play politics" within the military, thus undermining discipline and efficiency. With these conflicting attitudes to overcome, even among their fellow officers, it is little wonder that military leaders who ultimately do assume national leadership tend to be both tactically astute and politically tough.

The military's proposed solution to these conflicts, however, has generally been typically apolitical. Officers should be better trained. The Center for Higher Military Studies—whose directors have proudly proclaimed it to be not a "school for presidents" but definitely a "school for statesmen"—has not only improved military training, but it has retained the traditional military prescription for national health, adding only "technology." And the means by which to instill morality, discipline, patriotism, and technology remains one of the traditional panaceas of the Peruvian military: education. As with other military attitudes, the origins of this emphasis on education are to be sought in a mixture of institutional and social factors. The importance of education to the promotion process (a step that has revolutionized the military career since the 1930s) reflects, among other things, the concerns of men sensitized by the knowledge that low social standing and limited finances had, during their adolescence, precluded their attending civilian universities.

Even at the CAEM, however, politics in Peru, and especially good government, was still seen in the early 1960s as an imponderable fraught with difficulties and beset by devils. Bolívar's statement that he had "ploughed the sea" in trying to govern reflects a common military attitude. "So long as Peru does not have programmatic and well-organized political parties, *the country will continue to be ungovernable.*" [9] This ungovernability, however,

8. General José del Carmen Marín, who made this particular comment to the author in 1964, was referring to the 1930s primarily. By the 1960s and 1970s, partly because of Marín's efforts as an educator and more importantly because of the strengthening of the military as an institution, these roles were frequently reversed.
9. CAEM, *El Estado y la Política General,* Chorrillos, 1963, p. 89. Emphasis added.

was not attributed to the traditional and well-known civilian defects alone:

The sad and desperate truth is that in Peru, the real powers are not the Executive, the Legislative, the Judicial or the Electoral, but the *latifundists,* the exporters, the bankers, and the American [U.S.] investors.[10]

And the oligarchic and foreign devils are joined by the APRA party—"a form of national cancer," according to the same military planners.

Despite their hostility to politics and the tendency to oversimplify complicated issues for the sake of action, Peruvian military leaders have frequently demonstrated considerable flexibility and political skill. That some officers are capable of being *criollo* (clever and sometimes unscrupulous realists) does not alter their suspicion that political compromise is fundamentally a betrayal of military values, but it may enable them to put some of their theories to a rather effective test.

Practical Threat Perception: Guerrillas, Intelligence Organization, and Petroleum

The military education system, as we have seen, developed some interesting doctrines. It nonetheless took the guerrilla campaigns of 1965–1966 to force social theory out of the schools and into the barracks, thereby making the political immobility and economic decline of the late 1960s a matter of urgent military concern. In the summer of 1965, two separately organized guerrilla fronts opened in the central and southern Andes with ambushes of police units. Within a month, the outbreaks had led to the displacement of the relatively ineffective rural police by a joint military command and martial law in the affected forces. This in turn led to the discovery of other fronts still in the process of forming in other parts of the country. Within six months, despite forebodings in elite political circles about "revolution in the Andes," the military forces completely eliminated the guerrilla pockets and almost entirely wiped out the MIR leadership. And they did this without forcing a change in government, and without the prolonged suffering and mounting casualties characteristic of other cases of political violence (Guatemala and Colombia among them).[11]

Containment of the guerrilla threat also confirmed the military in their commitment to reform. The guerrillas had chosen for their headquarters a remote mountaintop called "Mesa Pelada" near the Convención Valley in

10. Ibid., p. 92.
11. The official account of the campaign is *Las guerrillas en el Perú y su represión* [The guerrillas in Peru and their repression], published by the Peruvian Ministry of War, Lima, 1966. See also Hector Béjar Rivera, *Perú 1965: Apuntes sobre una experiencia guerrillera* (Havana: Casa de las Americas, 1969), for an intelligent analysis by one of the few surviving leaders.

the Province of Cuzco, where the famous Trotskyist labor organizer, Hugo
Blanco, had successfully organized peasant unions in the early 1960s, before
his capture in 1963. But the Convención Valley had also been the scene of
construction of a penetration road from Colca to Amparaes by military
engineer battalions, and had been the site of a pilot agrarian reform program
by the military junta of 1962–1963. The failure of the region's peasants to
provide significant support to the insurgents appeared in military circles to
confirm the wisdom of the earlier reform policies.

The sense of success was tempered, however, by fear of a recrudescence
of violence. If a handful of radicalized urban intellectuals could occupy
thousands of troops for months, what would happen if popular forces and
the peasantry were enlisted in future disorders? The Ministry of War's published account of the guerrilla campaign concluded that Peru had entered a
period of "latent insurgency."

Nor was this a matter to be readily resolved with foreign assistance. Guerrilla war had proved the undoing of France, first in Indochina and then in
Algeria. French military operations had been observed by Peruvian officers
with French training and connections. Now Vietnam was proving the Achilles'
heel of the United States, demonstrating the difficulties that irregular warfare
could create even for the world's foremost military power. The conclusion
that the fate of these two historic military mentors seemed to suggest for
Peru was that internal subversion would have to be controlled by Peruvians
alone, if indeed it could be controlled at all.

The "latent insurgency" dilemma appeared to open many officers to the
idea that Peru needed agrarian reform combined with industrialization, or,
in the more abstract language of the Ministry of War, a "General Policy of
Economic and Social Development." [12] According to this view, similar to
the McNamara-Rostow thesis that violence springs from economic backwardness, conditions of injustice in the countryside needed to be removed
so that the absentee landowner and his local henchmen no longer would
exploit and oppress the rural peasant masses, whose marginal living conditions were making them potential recruits for future subversion and movements against military and governmental authorities.

Elimination of the latent state of subversion now became the primary
objective of military action. In a formal intelligence analysis by General
Mercado, the man who was to become Peru's foreign minister after the
1968 revolution, the "latent state of subversion" was defined as the presence
of Communist activity exploiting national weaknesses.[13] This Communist

12. *Las guerrillas en el Perú y su represión,* op. cit., p. 80.
13. General de Brigada E. P. Edgardo Mercado Jarrín, "Política y Estrategia Mili-
    tar en la Guerra Anti-Subversiva," *Revista Militar del Perú,* Chorrillos, No-
    vember–December 1967, pp. 4–33. An English language summary of this ar-
    ticle, "Insurgency in Latin America—Its Impact on Political and Military

activity, which took a variety of forms—military, political, economic, and social—was containable for the present. But the existence of national weaknesses continually threatened to point the balance against the forces of progress and order. National weak spots were defined, in General Mercado's remarkable statement of this theory, to cover a wide range of organizational, economic, technical, and political elements. His list of national weaknesses included fiscal crises, scarcity of trained personnel, resistance to change by privileged groups, inadequate scientific and technical development, lack of unity and coordination of efforts, absence of effective international security cooperation, lack of governmental control and communication with the rural areas, and, finally, lack of identification by the population with national political objectives. The reforms introduced by the revolutionary military government that took office less than a year after Mercado's article had appeared were largely meant to offset these weaknesses.

But threat perceptions, fear, and antisubversive warfare were not the only wellsprings of action. Genuine compassion for the conditions of the rural population was quite common among officers who had served in rural areas during regular tours as well as during the guerrilla campaign, and who often found emotional and ideological support for such feelings in paternalist Catholic social doctrines, which stressed that every man had a right to an existence offering material and spiritual dignity. In fact, there can be little doubt that among the major intellectual and moral forces impelling the largely Catholic military to action were the progressive priests and scholars who in the 1960s helped move the hierarchy of the Church in Peru to reorient its political participation in the direction of greater social justice for all Peruvians, including the poor.[14]

Communists, of whatever variety, were not, however, the only enemies of order and security in Peru. Many officers, accustomed since the War of the Pacific in 1879 to seeing external enemies exploit internal weaknesses, had come to believe that the United States, in alliance with Peru's "oligarchy," favored Peru's continuing in a state of underdevelopment. This view associated the United States with Peru's vulnerability to subversion as well as to more traditional external threats. The Belaúnde government had been strongly supported by most military men as a reformist movement dedicated to national progress. Its fumbling, which some attributed to American interests, only added to this theorizing, which to outsiders sometimes seems to verge on paranoia.

Strategy," by General Edgardo Mercado Jarrín, Peruvian Army, appeared in the *Military Review* published at Fort Leavenworth (March 1969, pp. 10–20).
14. See the discussion in Luigi Einaudi, Richard Maullin, Alfred Stepan, and Michael Fleet, *Latin American Institutional Development: The Changing Catholic Church* (The Rand Corporation, RM-6136-DOS, October 1969), especially pp. 51–55.

"Foreign interests, the oligarchy, and the decrepit politicians in their pay" was the way President Velasco was later to characterize this new subversive force, or *"anti-Patria."* [15] Although one result of studying political and social problems may be to realize their complexity, another may be to undermine the credibility of solutions advanced by political parties, thereby weakening the claim of civilian leadership to sole legitimacy. The legitimacy of civilian leadership was further eroded in Peru during the 1960s by the information collected through the increased activity of military intelligence services. The military command developed evidence of the corruption and compromises that were the daily fare of Peruvian politics. Even normal political compromise finds little acceptance in the military's values, as we have already discussed. Peruvian politics have never been very clean. Yet not every form of misconduct provokes indignation. A particularly messy smuggling scandal broke under the Belaúnde administration in early 1968. But smuggling, to military people, was sufficiently common not to be in itself cause to unseat Belaúnde. But in the context of the payment by private interests of "contributions" to political parties and leaders, of continuing inaction on basic reforms, and of congressional privilege, it was enough to lead increasing portions of the military, including nationalist elements in the intelligence services, to side with Catholic priests and others who denounced "corruption" in government.

All of these complicated matters were involved in the explosive petroleum issue that came to a head in the proposed Talara Agreement of 1968 and provided the immediate impetus behind the overthrow of Belaúnde and the installation of the government that still rules Peru today, two-and-a-half years later. This is not the place to review the petroleum affair,[16] except to point out that the debate over IPC became pivotal in helping to associate the United States government with the Peruvian and American private interests military officers were already increasingly perceiving to be inimical to Peruvian security and development.[17]

15. In a speech in Talara, 9 October 1969, commemorating the takeover of the International Petroleum Company. *Mensaje a la nación dirigido por el Señor General de División Presidente de la República, desde Talara, en el primer aniversario del día de la dignidad nacional* (Lima: Oficina Nacional de Información, 1969), p. 7.
16. Richard Goodwin described the intricate historical and legal background of the *La Brea y Pariñas* deposits, and the incredible ineptitude that marked the company's and Belaúnde's dealings with each other in his "Letter from Peru," *The New Yorker,* 17 May 1969.
17. Perhaps even more than the fishing industry (in which Peru now competes with Japan for world primacy), petroleum fits the category of a basic national resource. Contrary to some suspicions in the United States at the time that President Velasco had been irresponsible and unrepresentative, he almost certainly acted with broad military support in the IPC case. As early as 5 February 1960, the Joint Staff, over the signature of the Commanding General of

As General Arturo Cavero later explained to a group of visiting American military officers, threats to the internal security of Peru could originate in the plotting of groups opposed to peaceful revolution as well as in the efforts of groups who sought to impose revolution by violence.[18] General Cavero, who is now the director of CAEM, spoke when relations with the United States had improved from their low in early 1969, when application of the Hickenlooper Amendment seemed imminent, threatening to cut U.S. economic assistance and sugar quotas in response to the IPC nationalization. But Cavero began his remarks to the U.S. officers by pointedly quoting from General Mercado's speech before the United Nations in April of 1969:

The threat has varied over time. At first it was narrowly military in nature. Then new and subtler psychological and ideological threats arose against the security of each country. Today we face a new threat: economic aggression. Just as we fought against the violent aggression generated by guerrillas and by different forms of terrorism, so we are now fighting against economic aggression.

Cavero's U.S. military audience can have had no doubts about the direction of these remarks. Nor could they have had much doubt that the Peruvians meant what they said: In May 1969, less than a year before, Peru's military government had expelled the U.S. military missions from Peru.[19]

For a combination of reasons, then, many officers, particularly in the army, moved in the late 1960s toward an authoritarian preemption of what had traditionally been nationalist and left-wing positions, especially on petroleum and agrarian reform. But unlike the left, most of whose leaders dreamed of guerrillas or elections, the military were to impose their views

the Army, publicly recorded its belief that the *La Brea y Pariñas* agreements were "harmful to national sovereignty." That it took nearly nine years to put belief into practice is a sign of institutional caution rather than individual recklessness. An "inside dopester" account of the political history of the petroleum issue from the Peruvian perspective, of additional interest because shortly after writing it its author moved from the conservative newspaper *El Comercio* to the directorship of the Government Information Bureau run by the presidency, is Augusto Zimmerman Zavala, *La Historia Secreta del Petroleo,* Lima, 29 August 1968.

18. General Arturo Cavero Calixto, "Threats to the National Security of Peru and Their Implications for Hemispheric Security," a lecture delivered in Chorrillos at CAEM in April 1970.

19. The immediate cause of the expulsion was an attempt to retaliate for prior U.S. suspension of military sales (imposed in accordance with the requirements of the Pelly Act as a result of the perennial tuna disputes). The departure of the missions was also, however, delayed proof of the deterioration in both political and military relations between the two countries during the mid- and late 1960s. See Luigi R. Einaudi, *Peruvian Military Relations with the United States* (The Rand Corporation, P-4389, Santa Monica, California, June 1970).

under the aegis of a nationalist military dictatorship pledged to the non-violent restructuring of Peruvian society.

Revolution in Latin America in Light of the Peruvian Experience

The military government that has ruled Peru since October 1968 has clearly unleashed a process that is reshaping Peruvian society. Whether this process is called a "revolution" or not is, of course, partly a matter of perspective. For some, no process that is essentially nonviolent, or that leaves many traditional institutions formally intact, or that is led by military forces, or that is not led by a Communist party, can be called a revolution. And it is difficult to quarrel with such viewpoints, because the Peruvian government is led by the military, has avoided political repression, is willing to accept the Catholic Church and to defend private property, and has only the support, not the leadership, of the Communist Party.

But the real point is, of course, that while revolution can be *defined* in many ways, it is important not to lose sight of the social and political reality with which words are meant to deal. And the process under way in Peru today is clearly unprecedented. Parliament may return. Some future government may decide to pay Standard Oil an indemnity for the Talara oil fields formerly operated by the IPC and now administered by Petroperu. Some of the military ministers who have fought corruption may succumb to the temptations of power.

Yet should all this happen, Peru will still have changed in ways this writer considers decisive. The government has gone well beyond attempts that tinker with the status quo so as to "modernize" it or otherwise render it less objectionable. The agrarian reform begun in 1969 with the cooperativization of the productive coastal estates and the passage of the water rights law, and the subjection of new investment to the establishment of Worker's Communities (*comunidades laborales*) for profit sharing under the terms of the new industrial, fishing, and mining laws, are acts whose impact is more than legal.[20] They affect the fabric of Peruvian life in ways that cannot be reversed by changes in government or by new laws.

It is not yet entirely clear, of course, what that impact will be. It is also difficult therefore to identify who will be favored most by the process under way. If the middle classes and the foreign investors can take heart in the thought that the government does not in principle oppose the idea of private profit or property, the lower classes can also take heart in the government's

20. The purposes of these and other laws are detailed by the Chief of the Presidential Advisory Committee (COAP), General de Brigada E. P. José Graham Hurtado in his *Filosofía de la Revolución Peruana* (Lima, Peru: Oficina Nacional de Información, 14 de Abril de 1971).

willingness to recognize violent seizures of public lands for the purposes of private home construction by the landless poor and in the government's attempt to limit profits and ensure their equitable distribution.

Let us suppose, however, that much will go wrong, and also that some social sectors will profit considerably more than others in the changes under way. Let us even suppose that the one-third of Peru that is Indian in culture and life style profits least, directly, from what is now happening. Should we then argue that this is no revolution, or that this is just another military dictatorship to be opposed by all progressive and civilized people?

This writer would argue that three major points should be borne in mind in seeking the answer:

FIRST, what has already been done is a great deal, for Peru, and under present historical circumstances. Peru has not only one-third of its population unable to speak Spanish (and thus not easily mobilizable for national political—even revolutionary—purposes), but it has twice the population spread over ten times the territory with one-fifth the television sets per capita that Cuba has. Revolution in Peru, Cuban-style, would seem difficult even if a charismatic leader and an obligingly terrifying (and foolish) foreign enemy could be found.

SECOND, the changes that have taken place in Peru have come from within the system, not from outside it. Since the Cuban revolution, and especially since the rise of guerrilla movements, much attention has focused on "outside strategies" of change. Little hope was expressed for political leaders following "inside strategies" of change, that is, utilizing traditional institutions, like the military or the Church, for innovative political ends.

THIRD, we should not, however, assume that because of the relative success of the Peruvian military and the relative failure of Ché Guevara, that inside strategies will work where outside strategies have failed, or, alternatively, that the military have in the current decade become an inevitable force for progress. In the first place, the limits set by bureaucracy, internal divisions, and resource scarcity may in the long run produce a conservative reaction within the military even in Peru, where the process of military reorientation toward structural change has gone very far.[21] And finally, as our Peruvian case study has also made clear, those military intellectuals who argued for the "inside strategy" and for revolution from within, gained their audience among their fellow officers and comrades at arms not just because they were working within an institution with a highly developed educational system, but also because young civilian students and intellectuals gave their lives in following an outside strategy, and with their sacrifice gave the "insiders" a chance to make their views felt.

21. This is essentially the limiting argument I present in "The Military and Progress in the Third World," *Foreign Service Journal*, February 1971.

POSTSCRIPT, 1973

Five years after the October 1968 seizure, and two years after this essay
was originally written, the Peruvian revolution remains as difficult to define
as ever. In fact, much of its originality lies in its capacity to escape traditional formulas. President Velasco set the tone early, when he told his
fellow countrymen that

> "We can no longer afford to import foreign political or economic
> systems. We must develop, in a profound act of true creation, a social
> order that reflects the originality and history of our country and our
> continent." *

The search for such an order has led to a process of continuous experimentation in both agriculture and industry. New institutional arrangements
are being sought to lay the basis for a more just and modern nation. The
educational reform endorses bilingual and technical education. Both domestic
and international relationships are being quietly transformed. Peru is developing rapidly and has become a leader of the Andean Pact.

Throughout, the government has remained in military hands. The leaders
mentioned in the preceding pages are all still active. General Velasco,
despite a debilitating aortal aneurism that forced amputation of a leg early
in 1973, is still President. General Graham remains head of the Council of
Presidential Advisors. General Mercado has become Prime Minister and
Commanding General of the Army. General Cavero is now chief of the
joint military staff (*Comando Conjunto*).

But new leaders have emerged as well. Some generals are outstanding
technicians, like Generals Marco Fernández Baca, the brilliant president of
the new state oil agency Petroperú, Guillermo Marcó del Pont, chief of
the National Planning Institute, and Admiral Alberto Jiménez de Lucio,
Minister of Industry and Commerce. Others, like Generals Jorge Fernández
Maldonado, Minister of Mines, and Leonidas Rodríguez Figueroa, head of
SINAMOS, have considerable political skills as well. Some, like General
Francisco Morales Bermúdez, the Minister of Finance, are relatively young.
Others, like retired General Juan Bossio Collas, head of Mineroperú, are
older. Additional names, like Richter, Gallegos, Segura, Valdez and others
come instantly to mind. As General Graham once commented to me,
"behind us are new generations of officers, well trained, and ready to replace
us if we falter."

During the mid-1960s, none of President Belaúnde's nearly forty ministers
spoke Quechua. Today, several military ministers not only speak it, but are
not afraid to do so publicly. Few details are more indicative of the changes

* General Juan Velasco Alvarado, Message to the Nation, July 28, 1969.

of atmosphere over the past decade, and of the likely continuity of government committed to popular causes. But if the depth and quality of Peru's military rulers is impressive, so is their flexibility. Many different leadership combinations and specific policies are possible.

The notions of flexibility and time highlight two major uncertainties, however. Both are institutional. How great is the military's commitment *as an institution* to the revolutionary process now underway? What political institutions will the country ultimately adopt?

The evidence is unclear on both points. In the short run, certainly, major decisions will continue to depend on the internal workings of the military commands. Their nature will thus depend on the extent of shared experience and vision within the military leadership. In the longer run, of course, as Peru continues to develop, sectors other than the military will of necessity have to be brought into the decision-making process.

Nonetheless, the officer corps remains today the government's primary source of talent and support. But while many officers have been brought into the government, most, even at the higher ranks, remain outside. The routines of barracks life remain largely unchanged. Indeed, the very success of the current leadership in reorganizing government institutions and avoiding massive repression seems likely to deprive younger officers of the formative experiences and contact with the population that forged the revolution in the first place. Will schooling and example suffice to give the "new generations of officers" of whom Graham spoke a commitment to the revolutionary process begun in 1968? Or is the institutional tide within the military likely to shift toward more conservative and technocratic approaches?

Though the government has had substantial success in attracting able young civilian professionals to government service, the larger political question of civilian participation in leadership and policy choices also remains unresolved. SINAMOS, the government agency for popular mobilization founded in 1971, does not seem likely to provide a successful political base in the short run. Civilians who would prefer to participate in the revolutionary process, rather than merely serve it, still have few outlets for independent action.

Peru has come a long way. The road ahead will also be long. And new generations will have new revolutions to fight. Let us hope they will again be with, and not against, the military.

POLITICAL CLIENTELISM
AND CIVIL-MILITARY RELATIONS:
COLOMBIA

Steffen W. Schmidt

The military and military men have intervened directly in Colombian politics by seizing power only twice in the twentieth century. This pattern diverges dramatically from most Latin American countries and promises to be inconsistent with the political role of armies in Asia, Africa, and the Middle East as well. In the following pages we shall investigate the Colombian case, where the military seems on the surface to have been "a-political." This analysis will provide clues to understanding what conditions in the Colombian system account for this pattern of civil-military relations.[1]

I shall argue that one crucial difference between Colombia's and other systems has been the politicization of traditional, quasi-feudal patron-client (or landowner-peasant) relationships soon after independence from Spain.[2] By 1849 two distinct political clusters could already be identified. These groups became the nucleus of the Liberal and Conservative parties—still the only major parties in Colombia.[3]

This early intense split in the ruling class, and their success in mobilizing a following from the bottom up, prevented the growth of a highly integrated, unified oligarchy. Moreover, the mobilization of followers and the downward

1. Many studies of civil-military relations focus on systems where the army frequently seizes power or has become a semipermanent feature of politics. In Latin America this generally means Peru, Argentina, Brazil, and Bolivia.
2. I have argued this more extensively in "Bureaucrats as Modernizing Brokers? Clientelism in Colombia," *Comparative Politics,* forthcoming.
3. See German Colmenares, *Partidos Politicos y Clases Sociales* (Bogota: Universidad de Los Andes, 1968).

flow of political payoffs (i.e., the institutionalization of political clientelism) created the basis for high conflict in national politics, turnover in the presidency, and alternation of the party which ruled. Often this turnover was achieved by violent clashes pitting Liberal and Conservative armed partisans against one another.[4] Finally, one can demonstrate that the national army and police never came to develop political bases of their own and, especially in the case of the military, could never rule the country as a distinctly military institution.

Yet while the military in Colombia rarely ruled, the political parties have made use of this institution as a political resource. The armed forces have played a domestic political role as a supportive structure of the ruling party. This means that the military could not serve a single elite. The military did not become the tool of the upper class in its exploitation of the masses. The military in Colombia did not accelerate the stratification of society into horizontally conscious (class-based) strata.[5] Neither did it guarantee the rigid preservation of the status quo. How this configuration of forces related to each other and with what consequences to the system will be the focus of discussion in the following pages.[6]

The Growth of Clientelism and Political Identity

Patron-client relationships are the reciprocal exchange of goods, services, and loyalties. The literature suggests that these relationships lie at the base of Colombian politics.[7] The existence of these relationships throughout the political system had several consequences. First, it assured the growth of vertical loyalties over horizontal (class-based) ones.[8] The vertical relation-

4. The passion and "glory" of these groups is interestingly narrated by Gonzalo Paris Lozano in *Guerilleros del Tolima* (Manizales: Ed. Arturo Zapata, 1937). The author suggests that "war is the habitual state of humanity. It is the great purifier of people, a renovator *par excellence,* the greatest propellor of human destiny" (p. 99).
5. If any institution created obstacles to the acceleration of change and demands for change, it was the patron-client relationships themselves. One could even argue that highly elastic clientelism is a profoundly antirevolutionary force. It tends to adapt to new conditions, co-opt threatening elements (or eliminate them) and thereby reduce the tensions and consciousness leading to revolutionary demands.
6. For an interesting analysis of the dialectic between stability and change in Colombia, see Orlando Fals Borda, *La Subversion en Colombia* (Bogota: Tercer Mundo, 1967).
7. The classic socioanthropological insights into patron-client relations in Colombia is found in Orlando Fals Borda, *Peasant Society in the Colombian Andes* (Gainesville: University of Florida Press, 1962).
8. Douglas A. Chalmers' paper, "Parties and Society in Latin America," delivered at the 1968 Annual Meeting of the American Political Science Association, Washington, D.C., called my attention to the potential significance of patron-broker-client relations in Latin American politics.

ships were determined by the fact that the patron (boss) and his followers
were members of the same party. The vertical ties were further maintained
by the instrumental and affective needs of patrons and clients alike.[9] Thus
by the nineteenth century Colombia was a country where local notables of
the Conservative party mobilized their clients against local notables of the
Liberal party with their following and vice versa. Many nineteenth-century
politicians and most nineteenth-century Colombian presidents held military
titles achieved in these political wars. The national standing army, what little
of it there was, tended to be incorporated by the party in power. Since the
national army was partly composed of units scattered throughout the provinces, the loyalty of regional units often depended on which political party
dominated that particular region.[10] Therefore, if and when civil war between
the parties broke out, units in Liberal areas would tend to side with the Liberal cause and units in Conservative areas would side with that party. During
times of sustained conflict, presumably the party in power would purge opposition party sympathizers from the services altogether. Thus there was,
functionally speaking, no *national* army at all.

In addition the political parties could field armed partisans equal to and
often greater than the "national" army. Military rule was difficult simply
because the military did not have a monopoly over weapons and manpower
and because the national military never developed an identity separate and
distinct from the political parties (and primarily the ruling party).[11]

This fact overwhelms the Colombian picture and, in looking at three men
who tried to rule without large-scale political party support and who based
their power on the armed forces, we discover that they were unsuccessful.
General Jose Maria Melo in 1854, General Rafael Reyes in 1904, and General Gustavo Rojas Pinilla in 1953, all were swept out of office by mass opposition from the political parties. To rule in one's own name and repress

---

9. Patron-client relations are reciprocal ties. The patron gets from it farm labor,
   a loyal following, domestic servants, etc., while the client receives work, cash
   credit, legal help, even perhaps a sense of belonging.
10. Gonzalo Paris Lozano, op. cit. See also Jorge Martinez Landinez, *Historia
    Militar de Colombia,* vol. 1 (Bogota: Editorial Iqueima, 1956), for a thorough
    description of the civil war of 1899–1902 where the complexities of loyalties
    and military leadership become clear.
11. This remained true even after the establishment of military academies (1907)
    and the presumed professionalization of the armed services, which was ac-
    companied by more sophisticated weaponry. Guerrilla warfare waged by
    partisans continued to prevent the military from having a monopoly on the
    means of violence. To be sure, the military's greater sophistication contrib-
    uted to the increased casualties accompanying political conflict. In classic
    guerrilla fashion, most partisan bands armed themselves by attacking govern-
    ment troops or outposts and taking over the weapons and supplies of these.
    James L. Payne in *Patterns of Conflict in Colombia* (New Haven: Yale Uni-
    versity Press, 1968) has indicated that the Colombian standing army after
    independence was much smaller than that of neighboring countries.

the political parties solely through the use of the police and army has not been tolerated in this highly politicized system.

Recent Civil-Military Links

In conventional terms, the rather low profile of Colombia's armed forces in political life may warrant the label "nonpolitical military." [12] However we find a different use to which the army *was* put when we look at the relationship between the regime in power and the military in recent times. Here, the literature strongly suggests, politicians have regarded the police and military to be a "political resource" very similar to the rest of the bureaucracy. Clientelist systems are notoriously patronage-based (from patron-client relations). The civil bureaucracy appears to have been monopolized by the ruling party both in the staffing of the civil service (which is still almost entirely appointive) and in the preference given to sympathizers of the ruling party, in the disbursement of goods and services.[13] The military, on the other hand, was used to repress the opposition, facilitate the continued rule of the incumbent, and even guarantee that election outcomes would favor the party in power.[14]

One critical period in this respect was the Liberal party's coming into the presidency in 1930 after decades of being the minority opposition. This victory was possible because the ruling Conservatives were hopelessly divided in the late 1920s and ran two candidates in 1930, allowing the Liberals to win. The Liberals seized this opportunity to begin purging Conservatives out of offices, to take over the bureaucracy, and to terrorize Conservative sympathizers into silence. Violence between Liberals and Conservatives erupted, causing the government to crack down even harder on Conservatives.[15] In 1944 the abortive kidnapping by an army colonel of Liberal president Alfonso Lopez provided the opportunity to purge additional Conservatives out of the armed forces.[16]

12. Edwin Lieuwen places Colombia with Chile, Mexico, Uruguay, Costa Rica, and Bolivia. It is interesting to point out that for Bolivia what seemed like a nonpolitical army in 1961 bears no relation to the present. The Colombian case by Lieuwen's criteria still fits this picture although I am, of course, arguing that "nonpolitical" is an inappropriate term. *Arms and Politics in Latin America* (New York: Praeger, 1961).
13. See James L. Payne, op. cit., pp. 60–73; also, Fernando Guillen Martinez, *Raiz y Futuro de la Revolucion* (Bogota: Tercer Mundo, 1963).
14. Both parties have demonstrated this. For the Conservative case, see Pablo E. Valdez, *Historia del Conservatismo Colombiano* (Cali: Talleres "Renovacion," n.d.).
15. Abel Carbonell in *La Quincena Política* (Bogota: Ediciones Cosmos, 1952) brings together invaluable material on this period.
16. The kidnapper colonel (ret.), Diogenes Gil Mojica, in his memoirs disavows any political motive, claiming it was done in protest against Colombia's lack

Two spectacular assassinations of Conservatives in the 1930s by members of the armed forces underlined the extent to which the army was implicated with the Liberal government. On September 2, 1935, Dr. Juan Climaco Villegas, a prominent leader of the Conservative party, was killed by a Liberal army sergeant, Carlos Barrera Uribe, in the city of Manizales. Three years later on October 12, 1938, a Liberal army lieutenant, Jesus Cortes, assassinated Conservative journalist and politician Eduardo Galarza Ossa. These two incidents and the killing by police officers of a boxer nicknamed "Mamatoco," who apparently had inside knowledge of police corruption and of a murder perpetrated by the son of a prominent politician, stand out because they received national attention. Many thousands of cases in which Conservatives were persecuted by the military, police, and Liberal partisans have gone virtually undocumented.[17]

Two additional periods in the last twenty years can be identified as important. If we look at military expenditures (see figure I), two peaks can be detected which suggest an increasing level of importance of the military. In the illustration, the first peak grows from 1950 on and reaches its crest in 1955, or during the worst phase of a period of great political violence. Essentially the 1950s period constitutes a beefing up of the armed forces and a payoff to the officer corps by the Conservative administration of Laureano Gomez and the first years of the dictatorship of General Gustavo Rojas Pinilla. This buildup corresponds to the persecution of Liberal party members by the Conservative regime. The growth of military spending suggests that the military was viewed as a key supportive institution for the regime, a fact which also characterizes the early years of Rojas Pinilla's rule. Colombia's involvement in the Korean war may very well have been a move by President Gomez designed to extract larger amounts of military aid from the U.S. and gain support from the Colombian Armed Forces for doing so.

That is to say, Laureano Gomez and the Conservative party, much as the Liberals did in the 1930s and 1940s, could not easily have ruled from 1950 to 1953 without the implicit support of the military. The early 1950s saw a massive amount of violence erupt between followers of the two political parties. In many areas of the country this violence constituted large-scale insurgency, and in fact, the central government lost control over parts of the Eastern plains areas (the Llanos Orientales) which came to be ruled by local notables and guerrilla leaders, surrounded by armed partisans.[18] These Liberal enclaves were safe as long as they didn't threaten the centers of

---

of military preparedness for the short war with Peru (1932–33). *El 10 de Julio* (Bogota: Ed. Andes, 1971).

17. Valdez, op. cit.
18. A fascinating book on this subject is Eduardo Francisco Isaza, *Las Guerrillas del Llano* (Bogota: Libreria Mundial, 1959).

FIGURE I

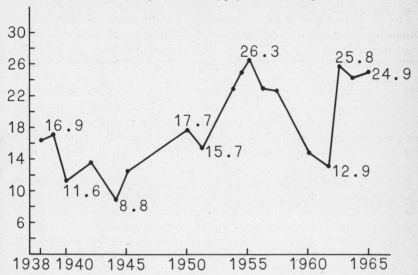

Colombian Military Expenditures as % of National Budget 1938–1965

Adapted from Joseph E. Loftus, *Latin American Defense Expenditures*, 1938–1965 (Santa Monica: The Rand Corporation, 1968), Tables 1 and 5.

power, the urban areas,[19] or as long as a status quo was maintained. The Colombian armed forces by and large constituted that guarantee.

At the same time, because of the explicitly political nature of the domestic violence, Liberals drafted into the armed forces during the Liberal administrations of Olaya, Lopez, and Santos in the 1930s and 1940s abandoned the armed forces in large numbers when the Conservatives came to power, often deserting to join Liberal party guerrilla groups. Thus one could conclude that the armed forces throughout the early 1950s came to be regarded by most Liberals as a Conservative force, there to defend political interests and not to be trusted. The case of Dumar Aljure, who in 1951 deserted the army, organized a sizable population on the Arari river area of the Department of Meta, and ruled as military commander and patron over a territory covering sixty thousand hectares, has many counterparts in this period.[20]

19. Only this fact accounts for the relative safety of the cities during even the worst violence which between 1949 and 1962 is said to have claimed about 250,000 lives.
20. Isaza, op. cit., and Richard Maullin, "The Fall of Dumar Aljure," *Rand* document 17010-1-ISA, 1968.

When General Rojas Pinilla overthrew Gomez in 1953, a second phase of military involvement began. In the first place the leadership of the Liberal party and factions in the Conservative, as well as a large portion of the military (including Rojas Pinilla), feared that Communists were taking advantage of the violence. Gomez was overthrown to prevent the violence from becoming revolutionary. Also there is evidence to suggest that during the Rojas Pinilla dictatorship, despite a short truce between government forces and guerrilla bands, official violence was unleashed with special ferocity. Villages were bombed, prisoners tortured, suspects executed by the army, and even urban political groups brutalized.[21] The most publicized of these was the "bull ring incident," in which hundreds were beaten and an indefinite number of people killed by the police, secret service, and other government forces. The episode took place as a result of the refusal of the spectators at a Bogota bullfight to pay respects to Rojas Pinilla and his entourage. Another incident was the massacre of nine students on the streets of Bogota during a demonstration. It should be mentioned in passing that after 1951 the police and military came under a joint command making it very difficult to discriminate between the two. It is clear that not only were both used as a highly partisan coercive arm of the ruling regime but also that both were perceived as such by the general population. Under the Rojas Pinilla regime the party-political identification of the armed forces tended to be less and the "personalist" use by the dictator in turn became greater.

The second sharp rise in military expenditures in figure 1 corresponds to a phase which can be called one of "pacification and counterinsurgency." [22] It was in this period that the first evidence of potential Marxist involvement in the rural violence appeared. Fidel Castro had toppled the Batista regime and soon took an aggressive stand on exporting revolution to Latin America. Colombia's Andes might easily have been the place for a second *Sierra Maestra*. However, by 1957 Rojas Pinilla had been ousted through the joint effort of the now reconciled Liberal and Conservative party leaders and by a segment of the armed forces which resented being dragged into an ever more complicated and messy job of internal repression. Thus by 1960, when the growth of the military budget once again began, the leaders of both parties and of the military had suppressed the urge to exterminate each other

21. A comprehensive overview of *la violencia* is given by German Guzman Campos et al., *La Violencia en Colombia* (Bogota: Tercer Mundo, 1964). See also Vernon L. Fluharty, *Dance of the Millions* (Pittsburgh: University of Pittsburgh Press, 1957), for an interesting discussion of the Rojas Pinilla regime.
22. For a discussion of the recent civil-military trends, especially counterinsurgency, see Francisco Leal Buitrago, "Politica e intervencion militar en Colombia," in Rodrigo Parra Sandoval, *La Dependencia Externa y el Desarrollo Politico de Colombia* (Bogota: Universidad Nacional, 1970) and Richard Maullin, *Soldiers, Guerrillas, and Politics in Colombia* (Lexington, Mass.: D. C. Heath and Company, 1973).

and had turned to face a potential threat to all of their interests—revolutionary insurgency and the possible radical transformation of Colombian society.[23]

The position of the armed forces in the 1960s has been underscored in the foreward written by Colonel Guillermo Plazas Olarte to a book about the violence of the late 1950s. He observes that, if it had not been for the Colombian army and men like the author of the book, a sergeant in the infantry, "the tricolor [flag of Colombia] left to us by Miranda would have been replaced by the hammer and sickle." [24] A former guerrilla leader, Arturo Alape, claimed that indeed a widespread incidence of revolutionary violence was under way during this period. He has written, "We must achieve the unity of the guerrillas where all are comrades, all are equal. This war is long or short but in the end it will be won by those who win over the masses." [25]

Through the late 1960s and early 1970s Colombia has been ruled under varying degrees of emergency powers, or in other words martial law, declared by the president so that he can cope with threats to law and order. The most obvious consequence of this suspension of civil rights under the constitution has been the use of military courts to try cases which are considered to threaten national security. From student strikes to oil workers riots, the Colombian army and police have become in the last decade an omnipresent fact of life. This period of "nonpolitical" military activity most clearly punctuates the inadequacy of conventional definitions of military intervention. I would argue that the Colombian military cannot be called a "nonpolitical" military.[26] The Colombian army is not ruling, but it is clearly a

---

23. The figures on U.S. military assistance to Colombia are quite interesting. From 1950 to 1963 Colombia received 39.4 million dollars, which averaged 3 million dollars per year. From 1964 to 1968 Colombia received 40.3 million dollars, which averaged 8 million dollars per year or almost a tripling of assistance. The figures for personnel trained under the Military Assistance Program indicate that between 1950 and 1963, 2,516 Colombians were trained, or an average of 194 per year, while the figure jumps to 276 per year in the five-year period 1964–1968. Adam Yarmolinsky, *The Military Establishment* (New York: Harper and Row, 1971), pp. 142 and 145, respectively.

24. Evelio Buitrago Salazar, *Zarpazo: Otra Cara de La Violencia* (Bogota: Imprenta de las Fuerzas Armadas, n.d.), p. 6.

25. Arturo Alape, *Diario de un Guerrillero* (Bogota: Ediciones Abejon Mono, 1970) p. 124. Perhaps the best book on the Latin American guerrilla experience written by a leader of the National Liberation Army is Jaime Arenas, *La Guerrilla por Dentro* (Bogota: Tercer Mundo, 1971). Arenas became disillusioned with the guerrillas, withdrew from them, and was assassinated on a Bogota street—presumably by members of the Marxist Left who argued that Arenas had betrayed the movement.

26. For instance in the early 1960s, Minister of War General Alberto Ruiz Novoa began to speak out about social issues, vigorously pursued "civic-military action" programs, and became so intensely involved in domestic politics that President Guillermo Leon Valencia dismissed him in January 1965 in the

critical pillar of the present regime's ruling structures, just as it has been to every regime thus far. It is for this reason that the "out" groups, especially, of course, radical Left elements, have attacked members of the armed forces verbally and in assassination attempts. As one student publication observed, "The army is the suit of armor without which the body of the oligarchy would long ago have been pierced by the arrows of the people's vanguard." [27]

Conclusion

The growing intensity of civil conflict in the 1950s and the interlude of the Rojas Pinilla dictatorship led in 1957 to a coalition between Liberals and Conservatives—the National Front. Under this arrangement the presidency alternates between the two parties every four years and all other positions, elected and appointed, are divided evenly between the two. The agreement was ratified by referendum, institutionalized as a constitutional amendment, and programmed to expire in the mid-1970s with a gradual renewal of competitive politics. One might suggest that from 1957 until the first years of the 1970s there has in fact been a single dominant elite—the leadership of the major factions in both parties.

This coalesced ruling group has been threatened by increasing student, labor, radical church, and peasant unrest. In response the military has been deployed to enforce "law and order"—to uphold the legitimacy of the National Front and hold the lid on. In the absence of political competition the two parties seem also to have neglected their clientelist networks. In effect, a guarantee of participation in the rule of Colombia since 1957 has devalued the patron-client basis of the party system. One of the consequences of this has been a rise of opposition political groups, which have been doing rather well at the polls and getting around the constitutional requirement of Liberal/Conservative sharing by prefacing their party name with the label of one of the traditional parties.

Has this interlude fundamentally transformed the relationship between the military and the sociopolitical leadership of the country? Will the "oligarchy," if there is one, for example, in economic terms, see their interest as coinciding more in the horizontal, class sense? Will they turn to the military as an alternative to the peasant and urban following which has in the past been the basis of their support? [28] Will the masses in turn become more

wake of rumors of a possible coup by Ruiz Novoa. See Leal Buitrago, op. cit., and also General Ruiz Novoas' book, *El Gran Desafío* (Bogota: Tercer Mundo, 1965).

27. From a mimeographed flyer distributed on the Universidad del Valle campus, Cali, Colombia, June 1972.

28. A parallel question can also be asked: Will the people come to look toward the military as a positive alternative to the existing civilian dominated type of regime?

isolated and neglected by the system and will they also come to develop greater horizontal identities; a more class-based sense of loyalty?

Scott has suggested that in Southeast Asia the accessibility to the landowner of colonial bureaucratic institutions has encouraged traditional patron-client networks to break down.[29] If one cannot only preserve but indeed enlarge one's wealth and power by having the army, police, and other national authorities guarantee one's status, one doesn't need or want the continued links with the masses, which are costly and time-consuming.

In the Colombian case the process has not run its full course. The National Front is scheduled to phase out. When this happens it seems probable that clientelist, competitive, politics will once again be activated. If clientelist politics is reactivated, the military is likely to represent a threatening, rather than desirable ally to the elite which finds itself in the opposition. A renewed differentiation of the two parties seems possible. If that is the case, civilian political consciousness is likely to continue, as it has during most of the past, to crowd the military out of the political center stage and relegate it to the role of supportive pillar to the party in power against the party out of power.

One clue is afforded by the responses of women surveyed in 1972. Only 30 percent said it might be necessary for the military to take over a government sometimes for "a period of time." Among the reasons given, the dominant answers included: "They bring law and order"; and "They should take over when there is anarchy." Others said that, under the military, "The country would progress"; "They would sweep out all the bad things"; and "There would be justice." An additional 30 percent had no opinion about the question of military rule while 40 percent were opposed to it. Among the dominant reasons for being against military rule were: "They shouldn't have anything to do with politics"; "They would deprive us of our freedom"; and "They shouldn't get involved in government." The data is from *Women's Political Attitudes Survey,* Cali, Colombia, July 1972, carried out by the author. The data from the 200 interviews using a 52 variable questionnaire is being analyzed and will be reported on in the near future.

29. Jim Scott, "The Erosion of Patron-Client Bonds and Social Change in Rural Southeast Asia," paper delivered at the Annual Meeting of the American Political Science Association, Washington, D.C., 1972.

THE POLICY-MAKING PROCESS IN A MILITARIZED REGIME: BRAZIL AFTER 1964*

Barry Ames

Since the 1964 coup, Brazil's military leaders have thoroughly dominated the political system. More than ever before in Latin America, they have expressed an intent to change the policy-making process, replacing personalism, nepotism, and corruption with "planning," "rationality," and the "public interest." Their success in turning this rhetoric into reality has not yet been evaluated.

This study focuses on changes made in the policy-making process following the coup.[1] It begins by discussing the post-coup elite and its goals—who they were and what they wanted to do. Problems in analyzing the implementation of the regime's rhetoric are confronted. Then, some hypotheses predict, under varying conditions, the degree of implementation of policy process goals. The hypotheses are tested in three policy areas: the removal

---

* Research for this article was supported in part by National Science Foundation Traineeship. I am indebted for comments and criticisms to Martin Needler, Paula Sornoff, and especially to Kitty MacKinnon.
1. In this study policy-making and decision-making will be distinguished by positing that decisions are choices between alternative solutions to single problems, while policies are collections of related decisions. In most cases (and in all the cases of this study), the language used by the participants themselves is indicative. A policy provides constraints guiding subsequent decisions. Each discrete decision may gradually or sharply change the broad policy. A policy-maker is an actor who is involved in a series of related high-level decisions. A decision-maker is a participant in a specific group. All policy-makers are decision-makers, but the reverse is not true.

of *favelas* (urban squatter settlements), the setting of national salary levels, and the allocation of federal educational resources.

Post-Coup Elite and Its Goals

At the time of the coup, three groups composed the anti-Goulart forces: (a) Castelo Branco and officers and civilian technicians associated with the Higher War School (the Escola Superior da Guerra or ESG), (b) civilian politicians, and (c) non-ESG officers.[2] The Castelo Branco regime soon became dominated by the ESG group. Civilian politicians like Adhemar de Barros and Carlos Lacerda lost their political influence. Generals sympathetic to the coup but not to the ESG were purged, including Olympio Mourão Filho and Amaury Kruel.

The ESG taught a systematic set of beliefs. Similar to the military in other developing countries, ESG ideology stressed nationalism, a puritanical outlook, acceptance of collective public enterprise, and a planning mentality based on an "antipolitics" position.[3] Some ESG publications argued that Brazil's political parties were too many and too localistic.[4] Others stressed the need for strong central government, economic planning, and mobilizing of resources.[5]

The ESG itself did not rule Brazil, but the ruling coalition of officers and civilians grounded their policy orientations in ESG doctrine. Their positions included short-term responses to the economic crisis, such as halting inflation and restoring economic growth, and long-term economic goals, such as maintaining private enterprise, a strong state role in the economy, and friendliness to foreign investment. Ideologically, they were strongly anti-Communist and sided with the U.S. in the cold war. Their institutional commitments included reducing party strife and eliminating "demagoguery" by populist politicians and union leaders. In policy-making, they were committed to authority and decision-making by planning and "rationality."

These orientations produced changes in policy input and in policy-making structures. On the input side, political parties were reduced to two. The political activity of thousands of people was repressed, and elections were devalued by reducing their number and restricting potential candidates' eligibility. Leaders of over four hundred labor unions were replaced, liberal

---

2. For background on the coup see Alfred Stepan, *The Military in Politics* (Princeton: Princeton University Press, 1971), pp. 240–41, and 176.
3. Cf. Morris Janowitz, *The Military in the Political Development of New Nations* (Chicago: The University of Chicago Press, 1964), pp. 63–65.
4. See David Carneiro, *Organização Política do Brasil,* Escola Superior da Guerra, Depto. de Estudos, c-47–59, pp. 18–20. Also see the *Revista Brasileira de Estudos Políticos,* no. 21 (Julho de 1966), a special issue on ESG doctrine.
5. Stepan, *Military in Politics,* pp. 178–82.

newspapers were attacked, and the right to strike was nearly eliminated. The regime attempted to repress the National Student Union (UNE) and prohibited marches and demonstrations. Changes in the policy-making structure included centralizing authority by decreasing the power of the legislature and the states and increasing the power of the executive, creating new ministries (e.g., the Ministry of Planning and General Coordination, to control budget design), removing authority to initiate new programs from certain ministries (e.g., the Ministry of Education and Culture), and attempting to improve bureaucratic performance by encouraging full-time work in government agencies.

Before evaluating the implementation of policy goals, the elite's concept of a "rationalized policy process" must be analyzed. Then the question of whether policy outcomes actually measure "true" goals can be confronted.

Elite informants were asked what they meant by a "rationalized policy process" and how they planned to improve it. Their vision of improved policy-making called for an increase in the number and authority of technically trained administrators (*técnicos*)[6] employed in decision-making posts. Elite informants expected *técnicos* to make decisions on technical criteria, but they did not expect perfectly rational and omniscient policymakers.[7] To assess their success, data was gathered on increases in the number of *técnicos* and on differences between their behavior and that of non-*técnicos*.

Since all regimes face constraints, using policy outcomes as tests of "true" goals[8] really asks whether the regime was powerful enough to realize its goals. Brazil may be a test case for this, since the regime was rarely forced to compromise its strongly held preferences. It crippled the unions, repressed student organizations, devalued elections, persecuted intellectuals, and emasculated the Congress. And although the regime faced financial limitations, its expectations did not overestimate Brazil's economic potential.

6. *"Técnicos"* in this analysis are administrators who have some expertise in their area of authority. They may be civilian or military. For a general discussion of Brazilian public administration, see Lawrence Graham, *Civil Service Reform in Brazil* (Austin: University of Texas Press, 1968).
7. For example, see D. Braybrooke and C. E. Lindblom, *A Strategy of Decision* (New York: The Free Press), part I.
8. Throughout this analysis the term "ideology" will be avoided. Discussion of authoritarian regimes often revolve around their ideologies or lack of ideologies, while tending to ignore the wide variations possible in regimes without comprehensive Nazi or Communist-type belief systems. All elites have goals. The real questions concern the nature of these goals, the intensity with which they are held, and the unity of beliefs among the elite.

Hypotheses

It is now well established that policy-making structures may vary according to the nature of the policy problem.[9] Variations can include the composition and goals of the policy-making elite, its need for supportive coalitions, the social groups making demands on policy-makers, the nature of policy-making structures during previous regimes, the amenability of the problem to "technical" solutions, and the experience of technically trained administrators in forming earlier policy.

While the regime's desire to improve the policy-making process was to affect all policy areas, many goals which related to specific groups pertained mainly to a single policy area. The priority the elite placed on "process" goals as opposed to "group" goals may differ among policy areas. This ranking will be affected by the aspects of the policy process mentioned above. Economic policy, for example, has not only been traditionally an area in which technically trained administrators wielded influence; but a general perception also exists that these issues are more amenable to technical inputs than, say, housing policy. Thus economic policy outcomes under a new military regime are more likely to reflect the impact of increased technical inputs.

All regimes develop clientele groups, and policy problems arise which affect them. Unless there is a complete harmony of interest between clienteles and the elite, the ability of clientele groups to influence policy ought to be inversely proportional to the unity and intensity of elite preferences on the issue. Technically trained administrators frequently must compete against clienteles of the regime. *Técnicos* are likely to be strongest when their competitors oppose the regime and weakest when their competitors support it. In addition, technical inputs should influence policy less when the policy elite is not united.

Though it is likely that the more technical expertise a decision-maker has, the more his decisional behavior will reflect it, expertise may make little difference. Available data are often spotty, inaccurate, and biased, encouraging decisions on other grounds. Moreover, data frequently are supplied by groups interested and involved in the policy process, so their biases tend to be reflected in decisions. The tendency of technically trained decision-makers to use their skills is also limited by the supportiveness of the decisional context. In milieus dominated by administrators lacking technical training, *técnicos* are less likely to behave differently from non-*técnicos*, because technically grounded decision-making may not be encouraged.[10]

9. Theodore Lowi, "Distribution, Regulation, Redistribution: The Functions of Government," in Randall Ripley, ed., *Public Policies and Their Politics* (New York: Norton, 1966).
10. See Anthony Downs, *Inside Bureaucracy* (Boston: Little, Brown, 1967), esp. p. 11.

Some factors influence outcomes more uniformly. Samuel Huntington points out that "prolonged military participation in politics inevitably means that the military reflects the divisions, stresses, and weaknesses of politics." [11] The longer the military is in power, the less characteristically military is its behavior, and the less independence it has from clientele groups.

Finally, the regime's success is limited by scarce technical and financial resources. If the regime concentrates technical personnel in one policy area, others will be shortchanged. In areas less important to the regime, policy-making patterns from earlier regimes are likely to persist.

A Methodological Note

The cases analyzed here are not a representative sample of Brazilian policy-making. They were chosen because they seemed important. In each policy area the analysis is developed chronologically. Within the narratives, data gathered in 1969 and 1970 from about sixty interviews with knowledgeable informants plus high- and low-level policy-makers are presented. These data reveal the behavior, attitudes, and perceptions of policy-makers.[12]

11. Samuel Huntington, *Changing Patterns of Military Politics* (New York: The Free Press, 1962), p. 36. Some recent writings have questioned the similarity of the goals of the various groups who have dominated post-1964 politics. The strengthening of the non-ESG "hard-line" wing of the military in the debacle following the elections of October 1965 have been called a "coup within a coup" by Ronald Schneider, and both Alfred Stepan and Cândido Mendes see a sharp contrast between the governments of Castelo Branco and Costa e Silva. In some areas of policy these analysts are clearly correct—both key policy-makers and their goals changed. The internal coup of 1965 did affect the composition and goals of the policy elite, but mainly in such areas as elections, parties, and purges. With the selection of Costa e Silva as the second president a gradual modification of economic and social policies began, including a "humanization of development" which supposedly meant a less repressive wage stabilization program and a more nationalistic attitude toward foreign investment. Moreover, the independence of the military elite from demands by nonmilitary groups which had supported the coup decreased during the Costa e Silva regime. Generally, I would argue that the discontinuity between the two regimes has been exaggerated and that many basic policies either were not modified under Costa e Silva or would have changed even if Castelo Branco and the ESG group had continued in power. Schneider suggests that Castelo agreed in 1966 to support Costa e Silva's bid for the presidency in return for a commitment to continue his program, and that after Castelo's death the regime followed a *more* Castelist line, especially in economic policy. This problem will be clearer after a discussion of the policy cases. See Ronald Schneider, *The Political System of Brazil* (New York: Columbia University Press, 1971), pp. 173 and 223; Cândido Mendes, "O governo Castelo Branco: paradigma e prognose," *Dados*, no. 2/3 (1967), especially pp. 107–11; and Stepan, *Military in Politics*, pp. 230–39.

12. Approximately half of each interview was open-ended, concentrating on the anecdotal details of the case, while the other half was structured, with memorized questions whose Portuguese translations had been verified by Brazilian

Analysis of key decisions combine with interview material to provide a clear, though impressionistic, picture of the *técnico* presence, influence, and decisional criteria.

Policy Toward Favelas in the State of Guanabara

To understand the changes in the policy-making process, old-style and new-style policy-making must be compared. Existing studies of pre-1964 policy-making are either too narrow or too superficial to allow rigorous comparison. But the policy-making process affecting the squatter settlements of Guanabara (in which Rio is located) was not moved to the federal level from the states and localities until 1968. Many important policy-makers had careers both before and after the coup. So *técnicos* and non-*técnicos* can be compared by examining the results of the 1968 federal preemption of *favela* policy. Further, the elite's relative indifference to *favela* policy permits examination of policy-making when the elite cares less about the results.

The General Context of Favela Policy

Containing anywhere from about one-half million to over one million residents,[13] Rio's *favelas* have long been seen as a problem by some government policy-makers. They have argued that *favelas* are unviable because their health conditions are bad, the hillsides threaten collapse during the rainy season, crime and promiscuity flourish, lands are desirable for nonhabitational uses, and they impair the city's attractiveness to tourists. *Favela* residents have long stated their preference for rehabilitating their homes rather than abandoning them.[14]

In the late forties and early fifties a few projects were developed, but since little money was available, the *favela* problem was largely "academic." In the early 1960s the Alliance for Progress and AID loaned Guanabara enough money to move about 6,000 families to such well-known projects as Vila Kennedy. But by the time of the coup only four *favelas* out of two hundred had been removed.

social scientists, especially Simon Schwartzman. The anonymity of respondents will be preserved. No tape recorder or written questionnaire was used. It should also be noted that while some officers were included, the majority of actual policy-makers interviewed were civilians.

13. The lowest estimate came from the Doxiadis firm in *Guanabara: A Plan for Urban Development* Comissão Executiva para o Desenvolvimento Urbano (CEDU), Doxiadis, Asso., Consultants on Development and Ekistics, Athens, 1965. For another estimate, see Rio: *Operacão Favela,* Governo do Estado da Guanabara, Rio, 1969, p. 25.
14. See Lawrence Salmen, "Urbanization and Development," in H. Jon Rosenbaum and William Tyler, *Contemporary Brazil: Issues in Economic and Political Development* (New York: Praeger, 1972), p. 429, for evidence on this point.

By 1964 the overall housing market in Brazil was in chaos because President Goulart refused to suspend rent control laws in the face of inflation. This reduced incentives for constructing all but luxury housing. The new government modified the rent control provisions and established the National Housing Bank (BNH) to invest directly in housing construction. The Bank was not primarily interested in low-income housing for *favela* residents, but it did agree with the Guanabara government to fund ten thousand low-income units, and by March 1966 about 8,000 families had been moved, mostly as a result of earlier projects. These projects were funded without significant habitational planning to decide which *favelas* should be removed and which should be urbanized. The banks knew that the residents' capacity to pay rent was doubtful. Why, then, were *favelas* removed?

Where sufficient information exists, it appears that criteria for removing the *favelas* often had little to do with living conditions. The *favela* Getúlio Vargas was removed to facilitate construction of the Rio-Santos highway. After furious, sometimes physical, confrontations between the residents and Governor Lacerda, part of Bras de Pina was demolished to make room for a factory.[15] The two largest *favelas* removed before 1966, Esqueleto and Pasmado, were also removed because of land claims. Esqueleto, built partially over water, was removed to provide land for the expansion of the University of Guanabara.

The residents' protests did not stop the removals. The policy-making environment changed, rather, because of increased information about recently built projects. By the end of 1965, it had become clear that there were serious problems in the big Alliance projects, Vila Aliança and Vila Kennedy. Nearly all the housing policy-makers interviewed saw that they were in bad repair, located too far from work centers, and cost the residents too much for transportation and rent.

In general the pre-1966 policy process was slow and halting, affecting perhaps four or five percent of the *favela* population.[16] Key elements influencing the process included limited resources for the construction of new residences, a general recognition of the enormity of the problem, lack of information, nonhousing claims on certain lands, and the residents' local electoral potential. Policy-makers responded first to nonhousing claims on *favela* lands, especially because sale of *favela* land provided capital to build units elsewhere. After the industrial, speculative, and intragovernmental

15. See *Jornal do Brasil,* 11 September, p. 9; 12 September, p. 13; *Correio da Manhã,* 21 October, p. 10; 23 December, p. 1.
16. It should also be noted that some limited urbanization projects were begun in those years. The huge *favela* of Jacarezinho in Jacarèpagua had been provided with water and sewage in 1965 at a cost of about one million dollars. Some smaller *favelas* had been aided by the Fundação Leão XIII beginning in 1962, and the *favela* Barro Vermelho had formed a cooperative in 1965 to build new houses and provide infrastructure.

claims had been met, the remaining units had to be filled with as little political stress as possible. *Favelas* selected for removal for genuinely habitational reasons tended to be the small and expendable ones. Smaller *favelas* were handicapped in fighting removal, having fewer resources and less organization, and because politicians had less to gain by supporting them.

A First Slice into the Policy Process: The Case of CEPE-3

The policy-making environment did not change until early 1966, when torrential rains inundated Rio, leaving 117 people dead and fifteen thousand homeless. Most of them were residents of *favelas,* the majority living on steep hillsides. Lacking strong foundations, many shacks simply collapsed in the loose soil.

The storms also unleashed a flood of statements, meetings, plans, and editorials. Even the most minor bureaucrat felt compelled to issue a position paper, and every position had some advocates. A host of study groups were formed, including groups representing universities, engineers, architects, *favela* residents, geologists, and social workers.

In December of 1966 the State of Guanabara formed an Executive Commission for Specific Projects (CEPE-3), mandated to design a definitive solution for the problem of the *favelas* of Rio.[17] CEPE-3's membership included the most important and knowledgeable men in housing from the federal, state, and private levels. Chaired by the secretary of the governor, it included representatives from government agencies involved in construction, and from the Geotechnic Institute, the state welfare agencies, the architects and engineers, and the construction industry. It also included Carlos Costa, who, besides being an influential relative of Costa e Silva, was a strong advocate of *favela* removal. CEPE-3 met over a period of five or six months, but it failed to reach an agreement, and the group was finally dissolved.

CEPE-3 was not merely a front group designed to take the heat off the governor. His desire for action was shown by the fact that he established the Urban Programming Office after CEPE-3 disbanded. CEPE-3 failed because of its political environment, and especially because its members feared that any actions they might take could have negative consequences, such as loss of jobs, for themselves.

CEPE-3 was scrutinized intensively in the research for this study. Ten of its fourteen members were interviewed, including men with and without technical skill and federal and nonfederal employees. Since many of the members had long bureaucratic experience, and since the federal government had not yet entered the *favela* policy arena, this group provides evidence of *técnico* behavior in a special political setting. Did *técnicos* use technical criteria in their decision-making behavior more often than non-*técnicos*? Did

17. This was the third such commission.

they respond to public rather than private interests? Did they attempt to plan solutions?

One way of illuminating differences in decision-making behavior is by examining the methods used to analyze and evaluate alternative proposals. The group's participants were questioned on their own goals relevant to the *favela* problem, the proposals they had made to attain these goals, the proposals made by other participants, and their reasons for accepting or rejecting these alternatives.

Even providing a generous definition of analysis, very little was done by CEPE-3. Group members frequently distorted the meaning of alternatives, and nearly all rejected various alternatives by "impossibility inferences;" i.e., unsupported statements asserting that a plan was impossible or impractical.[18] Programs were rejected as too clostly in the absence of cost data for any project. Frequently policy-makers rejected proposals on "technical" criteria which were outside their fields of competence. For example, one member gave his opinion "as a geologist" that all *favelas* in the South Zone (a mostly upper-class district) should be removed. His reasoning, that residential areas should be socioeconomically homogeneous, was clearly not based on geology.

Technical analysis was generally low, but it is probably more significant that participants with varying levels of technical skill and participants from both federal and nonfederal agencies did not differ among themselves in their decisional behavior. This raises the possibility that a crucial influence on decisional behavior may be the internal dynamic of the group itself, which, in this case, contributed to a disinterest in potentially feasible analysis.

One indicator that internal group dynamics were important was the unwillingness of most participants to support programs which might add to the activities of their own agencies. With a few exceptions they were not "Parkinsonian." One agency head stated that his unit had kept the same level of funding for the past few years and hoped to remain at that level. An interest group representative remarked that because his organization had been criticized for sending briefs to the government, it now wanted as little contact with the government as possible.

Why were the group's members so negative about new programs? Past experience had not been favorable. When group members were asked about the success of the large Alliance projects, they were highly negative, citing their still unfinished condition, high transportation costs, etc. The chances for success of any project were slim, so in a condition of high uncertainty it became rational to avoid real solutions.

18. An extensive analysis using this concept, derived from cognitive processing theory, is John Steinbrunner, "The Mind and Milieu of Policy Makers: A Case-Study of the MLF," (Ph.D. diss., Massachusetts Institute of Technology, 1968).

The search for information about the participants' interest in planning soon demonstrated that on the verbal level, at least, CEPE-3 was full of superplanners. Only two of the ten interviewees did not use some elements of the following language at least once:

A solution to the problem requires. . . . comprehensive planning. . . . attention to all aspects of the problem. . . . global planning, coordination of all sectors. . . . solution of transit, workplace, rents in one solution. . . . fusion of States of Rio and Guanabara so a comprehensive solution can be pursued. . . .

One of those not using this language had actively tried to recruit support for an experimental and only partially planned pilot urbanization project. Opposition to this plan was strongest from those who emphasized planning most. These data are too incomplete to permit final conclusions, but it may be that emphasis on "planning" (even if unrelated to any real planning) flows from and rationalizes a desire to avoid concrete commitments. "Planning" as an ideology may support and justify inaction.

In sum, the membership of CEPE-3 included *técnicos* in the truest sense: recognized experts in a technical subject bearing directly on the problem. The *técnicos'* methods of evaluation of proposals were not strikingly based on evidence. And if the "planning" mentality rationalized noncommitment, it worked for both *técnicos* and non-*técnicos*. Either these supposedly "new-style" *técnicos* do not really behave differently, or they behave differently only in certain decision-making contexts.

The Growth of Federal Power

One of the most significant long-term effects of the rains was the development of massive federal power in *favela* policy. State officials encouraged the federal government to take over, because the issue could only be a political liability for them. Moreover, federal officials decided that nothing would be accomplished as long as policy choice remained at the state level. The state could not formulate a comprehensive plan for *favela* eradication. It lacked sufficient resources to carry out such a plan, and the problem was really national in scope.

From early 1966 the possibility of creating a new organ to take over policy-making authority was discussed, especially in the Ministry of the Interior, whose chief, General Albuquerque Lima, was said to have been especially interested in *favela* policy.[19] The immediate stimulant to federal preemption was the announced decision of the state to remove all the *favelas*

19. For example, see an article by BNH Director Harry James Cole, in the *Revista de Administração Municipal* 79 (March and April 1966), p. 83.

in the South Zone. In late 1967 Vitor Pinheiro, head of social services in Guanabara, announced that all South Zone *favelas* would be transferred to the Centro Communitário Sul, to be built with funds obtained by selling the land of the *favelas* Praia do Pinto and Catacumba.

Official explanations for the move cited the unhealthy conditions of the *favelas,* but the real impetus came from a Guanabara real estate firm. Given the expansion of the BNH and the boom conditions in the housing market, the real estate firm expected large profits from buying the land and reselling it to developers of luxury high-rise apartments.[20] The plan was taken to Vitor Pinheiro, who considered himself against removal of *favela* residents. Pinheiro had a "pet project," the South Community Center. This "New Town" concept for Rio's poor would be separate from the wealth of the South Zone but much closer to work centers than Vila Kennedy or Cidade de Deus. Probably because it provided financing for his Centro, Pinheiro adopted and sponsored the real estate firm's plan.

The residents of Praia do Pinto, a large, well-organized, and politically strong *favela,* responded to the removal order by preparing an urbanization plan providing space for ten thousand residents plus the necessary infrastructure. They also sought support from architects and sociologists, and they were aided by the unwillingness of either of Rio's prestige newspapers to support the state's plan.

The plan seemed to be in jeopardy until the federal government stepped in. In May 1968 the Ministry of the Interior created CHISAM, the Coordination of Housing of Social Interest in the Metropolitan Area of Greater Rio. Directed by Gilberto Coufal of the BNH, CHISAM's staff included two engineers, two or three lawyers, an architect, two sociologists, a social worker and two or three secretaries. This small agency had authority to approve or veto all *favela* projects undertaken by all state-level agencies. Authority over *favela* policy had effectively left the state level.

While CHISAM had authority to remove bottlenecks in the policy process caused by the indecisive and often contradictory policies of state agencies, this did not insure a rationalized policy process. The federal government did not insulate CHISAM from pressures emanating from within the coup elite itself. The clashes between CHISAM and Carlos Costa, head of the governor's staff and brother-in-law of the president, illustrate the conflict between "new-style" and "old-style" bases of authority. CHISAM's technical studies did not conclude that all South Zone *favelas* had to be removed. The insistence on total removal came from Carlos Costa, whose influence at the ministerial level was greater. He also insisted even as late as July 1969 that by the 1970 elections all the *favelas* between the airport and the train

20. There is no documentary evidence for the statement that the impetus for this plan came from a real estate firm. The story was related independently by four different "trustworthy" informants and confirmed by another.

station area would be eradicated. CHISAM had no such plan and lacked resources to execute one. The influence of Carlos Costa in housing policy decreased only with the death of Costa e Silva.

Even if CHISAM's size and authority were to increase, the influence of technical criteria in policy-making might not. One reason is the paternalism of CHISAM officials, especially Director Coufal. Although recently CHISAM has blamed *favela* conditions in structural factors such as market inequities,[21] in interviews CHISAM directors and staff saw residents as socially undeveloped and needing character rehabilitation. Ending promiscuity and lifting the residents' spiritual level were frequently mentioned.

Often this paternalism was combined with a strong class bias. CHISAM Director Coufal did not believe that the dislocation of the project resident from his work locale was a serious problem:

The rich have problems too . . . everyone does . . . with the taxi or bus . . . everyone in Rio has problems with transportation . . . *favelados* are not a privileged class.[22]

Coufal was undoubtedly aware that residents of Vila Kennedy spend about twenty percent of their incomes and three to four hours a day on transportation. This callousness and indifference, coupled with a pervasive paternalism, supported the belief that all solutions must be planned from the top. Coupled with pressure from non-*técnico* influentials like Carlos Costa, these beliefs prevented the expansion of some promising urbanization projects begun by state agencies.

One such urbanization was located in Bras de Pina, a *favela* selected for self-urbanization because it appeared potentially viable. A plan was developed, providing a Guanabara agency, the Company for Community Development (CODESCO), to contract for the infrastructure, with loans to residents to build their own houses. Maximum loan repayments were between one-third and one-half the payments for apartments of comparable size in a nearby high-rise project.

CODESCO's original plan to begin an urbanization project every three months was not implemented, essentially because CODESCO lacked political support in the crucial funding agency, the National Housing Bank. The Bras de Pina project itself was apparently funded only because one high official in the BNH was favorable to urbanization and was also related to a member of the original study group. The CODESCO staff believed that the BNH opposition to the Bras de Pina project was based on a dislike of "improviza-

21. Cf. Josefina Albano, "O fator Humano nos Programas de Recuperação das Favelas," 1961 manuscript; and *CHISAM Origem-Objectivos. Programas-metas* (Rio: CHISAM, 1969).
22. Interview with Gilberto Coufal.

tional" planning and fear of mobilizing the residents politically. Since these projects were cheaper for both government and homeowner[23] and were obviously preferred by the residents, the growth of federal influence effectively closed off one of the most hopeful avenues of *favela* policy. Even on the regime's own criteria, federal influence did not lead to optimal policy outcomes.

Salary Policy After 1964

In salary policy the benefits of *técnico* decision-making should be most evident. The elite was united in its general ideological preferences, economic problems were salient at the time of the coup, and technically trained administrators had considerable influence even before 1964.

On the surface, salary policy does appear to conform to a "rational" model. Salary increases were regularly held below the rise in the cost of living, and the inflation rate dropped dramatically. However, it is questionable whether salary policy caused the reduction in the inflation rate. Albert Fishlow, a member of the U.S. AID Economic Mission to Brazil, argues that the dominant diagnosis of pre-1964 problems made by post-coup policy architects (especially Roberto Campos) led to excessive salary repression and an over-tightening of credit. This produced a decrease in investment and continuing high levels of inflation without corresponding growth. Increased growth between 1967 and 1970 resulted partly from the development of exports, but mainly from a business-cycle upswing.[24] Government policies according to Fishlow, were not primarily responsible for the economic reversal.

The 1964 coup unquestionably increased *técnico* presence in salary policy-making. To determine whether the presence of more *técnicos* improved the "quality" of decisions (given the goals of the elite) we will examine the views of high-level economic policy-makers, the general outlines of government wage policy, and a specific group faced with implementing an emergency wage increase in 1968.

The Coup and Economic Ideology

The economic climate of early 1964 was characterized by negative GNP growth, an annual inflation rate of 140 percent, and successful union strikes for large increases. Interviews and documentary evidence revealed that economic policy-makers generally agreed on economic goals. This is

23. Salmen cites figures estimating the cost of new housing at the equivalent of one thousand U.S. dollars per house, while servicing and upgrading *favelas* in present locations costs half that. Salmen, "Urbanization and Development," p. 430.
24. Speech at the Stanford-Berkeley Colloquium, 7 April 1971, Berkeley, California.

not surprising, given their common Higher War College background. Respondents (primarily connected with the Ministries of Labor, Treasury, and Planning) stressed the need to control the cost of living. Inflation was cited either first or second as the "most important problem" for every government policy-maker interviewed. Economic conditions in 1964 were so bad that it was unnecessary to consider the possibility that deflationary policy might be counterproductive for growth. A much simpler solution was to control inflation. A kind of "capitalism and free enterprise" value underlay respondents' answers to "What do you believe is the basic role of the state in economic development?" and "Do you believe that the major sectors of the economy now in private hands should remain there?" Representative answers included aiding free enterprise, helping foreign investment, and providing capital to national industry. Few respondents were interested in nationalizing any industry.

These economic policy-makers unanimously believed that the labor movement in 1964 had become Communist-dominated and "demagogic." They felt its leaders were calling for raises far in excess of cost-of-living increases and were thus fueling inflation. Some unions, such as the steelworkers, were perceived to be getting extraordinary raises simply because of their strike threat.[25]

Alternative weapons to salary restriction for fighting the inflation were rejected on a number of grounds. Asked why profits could not be restricted by profit-ceiling legislation, progressive taxation, or price control, some respondents argued that there was insufficient data to analyze profits, and/or that profits were not high anyway. Others thought that profit restriction was antithetical to free enterprise. Many believed that such moves would be self-defeating, because they would cut investment and restrict growth.

Policy-makers seem to have reasoned that forced saving by the lower class was necessary, and that only when nonmarket forces such as unions interfered with this classical pattern of capitalist development would the real wage of the salaried class rise and become inflationary. Roberto Campos

25. An economist's perspective on the wage situation would differ somewhat. According to Peter Gregory's study of industrial wage policy between 1959 and 1967, real wages rose by 4.1 percent per year between 1959 and 1962, and by seven percent per year in 1963 and 1964. The 1959–62 period was one of expansion in which market forces produced a rise in real wages, because many categories of skilled labor were in short supply. The 1962–64 period was characterized by high inflation, lessened growth, and a very loose labor supply, so the rise in the real wage must have been a function of such nonmarket forces as unions or government. The real losers in this period were not the industrialists, who were able to defend their profit position, but property owners and agriculturalists. In urban areas rents were largely fixed, and the government kept basic food prices low as a subsidy to urban workers. See Peter Gregory, "Evolution of Industrial Wage Policy in Brazil, 1959–1967," USAID/Brazil Summer Research Program, September 1968.

explained his unwillingness to control inflation through freezing prices by arguing that

It would be counterproductive: it would impede the readjustment of agricultural prices, essential for the modernization of agriculture; it would congeal the prices of basic services, diminishing the capacity of investment in energy and transportation, and finally, it would lend itself to fraud and abuse, because of the inefficiency of the administrative machinery and the inadequate moral discipline of our controls.[26]

The policy elite also wanted a *definite* policy. When asked "What have been the major achievements of the salary policy?" about one-third of those interviewed mentioned the need for *one* policy. Others concurred spontaneously; the remainder agreed when asked if they thought having just one policy was a virtue in itself.[27]

The Development of Government Wage Policy

Although the conspirators took office without fully developed economic plans, the unions received quick attention. Forty labor leaders were replaced within the first month after the coup. By the end of May, the government admitted having intervened in three hundred unions, including almost all the important national ones. A comprehensive salary policy emerged slowly. The National Council of Salary Policy was reorganized, and the government declared its intent to maintain the working-class share of the national wealth. The current share to be maintained would be the average salary for the previous twenty-four months. In 1964 this formula was applied almost literally, causing a sharp decline in real wages, because, with continuing inflation, only in the first month after the increase did a worker's salary actually equal the twenty-four-month mean.

In 1964 government actions were supported by all sectors of business and by all major newspapers. The workers' organizations still functioned, but they lacked power to influence government policy. By 1965, however, the unanimity of support from business and media began to break down, and splits appeared within the government as well.

The business community divided over wage policy because commercial interests began to suffer from the decline in middle- and lower-class buying

26. *Correio da Manhã*, 5 August 1966, p. 11. Campos was speaking to the General Staff of the Armed Forces. When economist Mário H. Simonsen opposed the emergency salary raise of 1968, he commented that workers already received beyond what is fitting to the national economy," *Correio da Manhã*, 27 April 1968, p. 3.
27. For an illustration of this terminology, see an interview with Ivo Pinheiro, Director of the National Salary Department and a respondent in this study, in *Visão*, 25 October 1968, p. 49.

power which resulted from wage repression. This split was reflected in the government itself. For example, a faction of the National Economic Council (CNE) led by Fernando Gasparian began attacking government policy as monetarist and orthodox.[28] The criticism leveled by Gasparian was discounted by some policy-makers on the grounds that his textile family background biased his views, because textiles in Brazil were largely obsolete technologically and were among the first to feel buying power declines.[29] By July 1965 *Correio da Manhã* was talking about "serious" unemployment in textiles and arguing that government tightening of credit was discouraging modernization. Gasparian was soon joined by another "highly placed" critic, Antônio Dias Leite, an economics professor from São Paulo. Their anti-monetarist criticisms began appearing in trade journals such as *Digesto Econômico,* the journal of the Commercial Associations of São Paulo. Their attempts to form a political alliance with presidential hopeful Carlos Lacerda were fruitless, since he was not allowed to seek public office; therefore this essentially in-house protest of the CNE and its allies yielded nothing.

In May of 1965 the first formal plan was published by the Ministry of Planning and General Coordination. Known as the "Program of Economic Action of the Government 1964–66," it established guidelines for wage policy:

The salary readjustment will be determined so as to equal the real average salary for the last 24 months, multiplied by a coefficient which measures the increase in productivity estimated for the previous year, plus a provision to compensate for the chance inflationary residual allowed in the financial programming of the Government.[30]

28. The CNE is a kind of free-advice commission which united representatives of major economic sectors: labor, manufacturing, commerce, etc., with *técnicos* from various ministries.
29. Interview data.
30. *Programa de Ação Econômica do Governo 1964–66,* 2d ed., Documenta EPEA, no. 1, May 1965, p. 85.
    The first formula for wage increases can be expressed mathematically as follows:

$$W_0 = \frac{\sum_{i=1}^{24} \frac{wi}{ci} \overline{\frac{(R)}{(1+2+p)}}}{24} \quad \text{where}$$

W_0 = newly adjusted wage level.
W_i = the money wage in force during the preceding 24 months.
C_i = a price index with a base equal to 1.00 in month 1, the last month prior to the effective

The inflationary residual was designed to compensate workers for the continuous decline in their buying power each month after their annual wage increase by including a prediction of the inflation in the year to come. In 1966, for example, the predicted inflation was 10 percent, which means that salaries were increased by 5 percent extra (one-half the prediction). The increase in the cost of living between July 1965 and July 1966 was 40.5 percent. The next prediction, August 1966 to August 1967, was again 10 percent. The real cost-of-living increase was 30.1 percent.[31] In both years the real value of salaries dropped due to the over-optimism of the government prediction.

Did high-level decision-makers actually believe that inflation could be kept to ten percent? Or were they simply striving to achieve the maximum deflationary effect with the minimum political difficulty? Journalistic discussions presumed that the economic policy-makers were simply over-optimistic, that they were making honest errors. But the *técnicos* in the Ministries of Planning and Treasury admitted that no one really believed inflation could be kept to ten percent either in 1966 or 1967. Ten percent was picked as an optimum goal. If salaries were adjusted on that basis, they clearly would not contribute to inflation. If salaries were adjusted at some "likely" level, then a rise in the cost of living in that range might be accelerated by salary increases. It was safe to underestimate the prediction—it might help and it could not hurt. High-level policy-makers pursued this line of reasoning because their true objective was to restrain inflation at any cost. Maintaining the workers' share of the national product was official policy, but it was a goal of lower priority.[32] In this case deficient technical expertise, i.e., the inability to predict accurately the inflation rate for the next year, led to a salary repression even beyond the intentions of government policy-makers, although it was acceptable to them.

In 1967 the Costa e Silva government assumed power. Bulhões and Campos were replaced by Delfim Neto, a former economist at the University of São Paulo, and by Hélio Beltrão, who was much less competent and

---

month of the new adjustment
$R =$ expected rate of inflation over
the forthcoming twelve months
$P =$ national average of increases
in productivity.

31. These data are from Gregory, "Evaluation of Industrial Wage Policy in Brazil," table 2.
32. As Peter Gregory rightly notes, Campos and Bulhões believed that staple foods and utilities were unnaturally subsidized before 1964, and therefore these price increases should not "count" in figuring the new cost of living. See, for example, the *Boletim* of the Ministry of Labor, January and December, 1966, for an article by Francisco de Paula de Castro Lima, head of the National Salary Department and an appointee of Campos, for an exposition of this viewpoint.

less influential on policy. The attack on government policy by the commercial sector continued, and almost immediately there was a period of confessions inside the government. Everyone admitted that the salary policy of Campos and Bulhões had injured the workers. In May, Jarbas Passarinho, the new Minister of Labor, charged the National Council on Salary Policy with the task of bringing in suggestions for changes in the policy, adding that if the problem were not resolved by September 1, he would resign.[33] Unfortunately for the labor minister, the post-coup policy-making pattern had relegated his ministry to a secondary role. Passarinho's announcement on August 31 that the current salary policy would remain in effect until inflation dropped to ten percent (a figure it has not even approached) indicated that the monetarism of Delfim Neto prevailed with Costa e Silva.[34]

In September of 1967 the inflationary residual was increased to fifteen percent, meaning that salaries would be 7.5 percent higher to account for the next period's inflation. In the next year the cost of living rose 21.5 percent. So, although real salaries continued to drop, they dropped at a lesser rate. Salary increases in both the civil servant and minimum salary categories for 1968 were pegged at about twenty percent, much closer to the expected inflation.

A Second Slice into the Policy Process: The Emergency Bonus of 1968

In spite of Passarinho's "efforts," opposition continued to grow in Congress. In the Chamber of Deputies forty members formed the Anti-Salary Strangulation Parliamentary Front, composed of twenty-eight MDB deputies and twelve from the official government party, the National Renovating Alliance (ARENA). Senator Carvalho Pinto, a *paulista* ARENA member and a pre-coup finance minister, introduced a bill granting a ten percent advance to all workers who had not received one within the last six months. The advance would be financed mostly from excess welfare funds, but partially (one-fourth) by the employers. This bill was conceded a good chance to pass the Congress, and even if vetoed, it would embarrass the government.

In April 1968 Passarinho proposed his own version of the *abono de emergência* (emergency advance) which was similar to that of Carvalho Pinto. Unfortunately, the labor minister had not cleared the plan with the other ministers, especially those of treasury and planning. Apparently he had received general approval from Costa e Silva, but, given Costa's penchant for decentralizing authority to lower officials, this did not insure unwavering support. Minister Delfim Neto criticized the *abono* as inflationary; Housing Bank officials complained about a potential cutback in

33. *CM,* 9 May 1967, p. 5.
34. See *CM,* 31 August 1967, p. 5.

their fund supply. It was also discovered that treasury officials had been using the unemployment fund, from which the *abono* would largely be financed, to cover short-term federal budget deficits.

Manufacturers strongly opposed the plan. Heads of various industrial confederations immediately denounced the *abono* as demagogic and inflationary.[35] Their discovery that the plan had not been cleared with other ministries was used as evidence that it was ill-planned. The manufacturers' pressure was directed at the Ministers of Treasury and Planning. They in turn protested to Costa e Silva, who decided that Passarinho was being too hasty. A five-man committee (two representatives from the Ministries of Labor and Treasury and one from the Planning Ministry) was formed to study the *abono* and make recommendations.

The "official" mandate of the group was to design a noninflationary solution to the *abono* dilemma acceptable to both Delfim and Passarinho. After an intensive week-long session, such a plan, differing considerably from the Passarinho proposal, was developed. Recipients of the minimum salary (about one-half of the labor force) were excluded from its benefits, and no one could receive a payment of more than about one-third the minimum (twelve dollars). Rather than the unemployment fund financing most of the *abono,* it would be paid for by the Welfare Institutes. The government also agreed to cut back a part of a projected increase in the Merchandise Consumption Tax (ICM).[36] This rollback enabled some businesses to profit from the *abono,* since the difference in the ICM was greater than the payout the *abono* required.

When the decision group agreed, and the ministers accepted the plan, it was proposed to representatives of business. In a ceremony in the Treasury Ministry in Rio, the heads of the CNI and the Federations of Industry of Rio and São Paulo signed an agreement not to raise prices as a result of the *abono*. But its force was purely moral.

The bureaucratic group which designed the *abono* was selected for intensive analysis, including interviews with all members, because it seemed typical of the kind of decision-making the regime wanted to institute. The members were all technically trained, had worked together, and professed mutual respect (in contrast to CEPE-3 members). Within the broad constraints imposed by the ministers, the group had considerable power, because the potential costs to the ministers of deciding the issue themselves

---

35. *Jornal do Brasil,* 24 April 1968, p. 4. The newspaper attacked the *abono* editorially on 26 April 1968, p. 6, branding it opportunistic and suggesting that the Ministry of Planning was doubtful on it and that the Treasury Ministry was in the dark. See also *Correio da Manhã,* 25 April 1968, p. 10.
36. The ICM is a state tax, but its level is set by the federal government following negotiations with the states. It was regarded by businessmen as one of the most oppressive taxes.

were high, and a plan devised by a respected group of *técnicos* would have the legitimacy to insure acceptance.

The techniques for evaluating proposals differed sharply from those of CEPE-3. Proposals were not distorted; these *técnicos* understood each other's ideas and language. Alternatives were not dismissed as "impossible" without analysis. Group members did not apply cost-benefit analysis to every proposal, but they avoided intense involvements in their own proposals and concomitant distortion of alternatives.

Since even the *técnicos* on CEPE-3 did not display a noticeably technical decisional style, the difference in decisional style cannot be explained by the *técnico* composition of the *abono* group. The most striking differences lie in the internal dynamics of the two groups. In contrast to CEPE-3, the *abono* group had to reach agreement (the ministers insisted). They agreed on goals, and they liked each other personally. When an issue remained in dispute even after long discussion, the group compromised. All the members supported the final solution, even though aspects of it contravened the wishes of the superiors of individual members. Such actions would have been impossible for CEPE-3.

Members of the CEPE-3 group seemed to use the language of comprehensive planning as a rationale for inaction. In the *abono* group talk of total planning was much less often heard. Members seemed more concerned with particular bottlenecks, and they used terms like "flexible" and "pragmatic" instead of "integrated" or "comprehensive." [37] Perhaps part of the difference can be explained by the problems faced by the two groups. Certainly a one-month salary increase is simpler than a solution for the problems of Rio's *favelas*. However, when the *abono* group members were asked about their subsequent participation in a commission to design a new salary policy, they continued to avoid terms such as "comprehensive planning." Perhaps their incentives for action discouraged the use of terms which appear to hinder task fulfillment.

The increased effectiveness of decision-making does not mean that decisions were made on narrowly technical grounds. In fact, key choices were made to satisfy particular influential people or groups whose approval was needed. The *abono* was financed by the Welfare Institutes rather than by the unemployment compensation fund not because it was economically better, but because this choice reduced opposition from inside the government. The ICM was reduced one percent to provide a "carrot" to business. Even if the members had wanted to perform more technical analysis, it would have been difficult, because time and data were limited. This techni-

---

37. Albert Hirschman, in *Journeys Toward Progress* (New York: Anchor, 1965), uses these terms to describe opposite but equally laudatory approaches to policy-making.

cal weakness led to a bizarre outcome: a supposed emergency raise for
workers finally added to business profits because the tax break was bigger
than the payout. Since the task of this group was to avoid inflation, both
weak and strong firms had to be protected. Since the weak firms could
not be identified and exempted, *all* firms had to be protected, thus giving
the stronger firms a break. So in this case the decision-makers' inability
to use technical criteria prevented changing government wage policy even
when policy-makers knew it needed it.

Allocation of Educational Resources

At the time of the coup, the policy elite did not have strong and common preferences about education, but elite disinterest in the problem could
not keep it from becoming a salient public issue. In 1967 and 1968 the
issue of level and division of educational resources became the focus of a
crisis. Lack of unity and interest by elites in educational issues encouraged
particularistic clientele groups to participate. By this time the policy process
affecting education had been substantially modified; it was, in the perception of elite policy-makers, "rationalized." There were many more
técnicos as a result of changes in the policy process, but they faced many
competitors in their search for influence. By analyzing budgetary outcomes
for education, we will consider the causes of the changes that occurred
in the structure of policy-making, the changes themselves, and their impact
on the division of resources in education by university, secondary, and
primary levels during the Costa e Silva regime.

In spite of the elite's disinterest in education, the first budget of the new
regime (1965) included a sharp increase for the Ministry of Education
and Culture (MEC). In one year the federal budget's percentage for education rose from 5.1 percent to 7.5 percent.[38] Policy-makers in the Ministry of Planning and in MEC suggested that spending increased because other
ministries were not prepared to begin new projects and because education
was politically popular.

Unfortunately, MEC was extremely wasteful. The large influx of new
funds, in the opinion of informants, did not increase educational output.[39]
MEC's unwise use of funds had two results. First, since 1965 real spending
has declined every year at the primary school level and, until 1968, also
at the secondary school level.[40] Second, the elite's disenchantment with

38. Institute of Applied Economic Research (IPEA), "Recursos Públicos Apli-
     cados em Educação—1960–1967" (Rio: Min. do Planejamento, 1968), pp.
     16–17.
39. Interviews with director of Subsecretariat of Budget and Finance (SOF) of
     Planning Ministry and *técnicos* in IPEA.
40. For 1966–68 figures, see the *Anuário Estatístico do Brasil* for 1969 (Rio:

MEC encouraged attempts to improve educational policy-making. Professional educators, located mainly in MEC, began to lose authority not only over the total educational allotment, but also over the sectoral distribution of funds and even over specific educational policies, such as determining the location of new universities. Their loss of control resulted both from the growth in authority of the Planning Ministry and its high-level administrators' belief that educational policy-making could not be left to MEC.

Roberto Campos was a prime mover in the Administrative Reform of 1967 which sharply limited the power of the Education Ministry.[41] Campos established the Secretariat of Budget and Finance (SOF) in the Ministry of Planning, formalizing that ministry's control over each department dealing with spending. The Secretariat had a subsecretary for health and education who met with the Secretary-General of MEC and the head of each department to decide the budget for that department. Campos's innovation—changing sectoral budget-making from Treasury to Planning—supposedly added planning criteria to purely financial criteria in the educational budget process.

Also under Campos's aegis, the Ministry of Planning formed an educational planning agency which was to become a countervailing power to MEC. Currently called the Institute of Applied Economic Research (IPEA), its *técnicos* are experts in manpower planning and resource allocation. It must approve any project originating in MEC which calls for spending additional funds.

The education *técnicos*' publications were ambiguous as to which level of education they thought most needed new funding. But in interviews, members of the staff of the Human Resources Center of IPEA emphasized the need for priority for secondary education. Although more funds were needed at all levels, university-level spending was thought to lead to underemployment and unemployment among professionals. Most important, they

---

Instituto Brasileiro de Geografia e Estatística, 1969), p. 605. For 1965 the figures are from the *Anuário* of 1968, p. 503; for 1964, from the *Anuário* of 1965, p. 393.

41. Campos wrote in 1968, "The Law of Directives and Bases revealed the unrealism of attributing to the Federal Council on Education (an attribution happily removed by the recent administrative reform) the responsibility of planning the Federal system of education. . . . The planning of education cannot be the exclusive job of *educators* and *men of letters* (his italics) because it involves the fixing of priorities in the apportionment of resources . . . the analysis of the labor market and the calculation of costs, which is the job of the economist, as well as the organization of the logistics of the university, jobs for *técnicos* of education." *Estado de São Paulo*, 2 June 1968, quoted by Lauro de Oliveira Lima, *O Impasse na Eucação* (Rio: Editora Vozes, Ltd., 1969), p. 100. This book is the best general source of data and information on Brazilian education.

believed that Brazil's economic expansion required more people with secondary school-level technical skills.

The education *técnicos* faced considerable competition in policy-making. The Council of Rectors, representing the university administrators, lobbied forcefully for more university-level spending. And the planning minister, who one might expect to be a major advocate of the *técnico* position, was seen, with some influential economists around him, as a kind of Herman Kahnian "hard-headed realist." The education *técnicos* would become incensed at remarks such as economist Mário H. Simonsen's "as Roberto Campos says, Brazil does not spend absurdly little in education, it spends absurdly badly." [42]

Possible allies for the IPEA *técnicos* would be educational planners and some department heads based in MEC. They could well be regarded as *técnicos* of education since they had expertise and long experience in the area. But there seemed to be considerable hostility between the two groups, probably caused by the shift in authority from MEC to the new planners. The IPEA staff criticized MEC as incapable of carrying out anything, full of political appointees and incompetents, and dedicated to self-aggrandizement at the expense of programs. MEC bureaucrats thought *técnicos* were interested in quantity rather than quality and were unwilling to listen to people who had spent their lives in the field.

Another natural ally of *técnicos* seeking expansion of secondary education would be parents' and teachers' organizations. However, the activity of such organizations was generally inhibited by the repression of groups articulating lower-class interests. Teachers' organizations had to spend most of their time trying to get their members paid. No groups could be found whose primary purpose lay in expanding educational opportunities for poor rural children.

The structure of the budgetary process encouraged demand articulation by representatives of the universities, whose beneficiaries are mainly upper-middle and upper-class groups. The budget for each MEC unit was negotiated between a representative of the Ministry of Planning and the Secretary-General of MEC. To understand the process, one must know what a "unit" is. MEC is a series of departments, some with executive responsibility for their areas and some without. For example, the Department of Higher Education did not have responsibility for the universities. Instead, each university proposed its own budget to the Secretary-General and then negotiated the budget with the Planning Ministry. This gave the universities a considerable advantage, because it was much easier to present a complete informational case for a small unit, and because the rector could easily use his personal relationship with the president as a resource. Secondary

42. Mário H. Simonsen, Jr., "O Problema Educacional," *Industria e Produtividade,* April 1968, p. 8.

and primary education had to negotiate as overall categories instead of as individual entities.[43]

Shifts in Allocation Under Costa e Silva

Dissatisfaction with the results of the 1965 spending increase led to a continuous decline in spending for primary education. After 1965, secondary and university spending continued to record huge increases, but in 1965 and 1966 secondary education gained more than higher education, while in 1967 and 1968 higher education reversed that pattern, gaining much more in both years. These data are shown in table 1.[44]

Table 1: *Federal Expenditures by Level of Education 1964–1968*
(in 1000 current NCr$)

	Primary	Index	Secondary	Index	University	Index	Price Index
1964	25,748	100	34,180	100	80,500	100	100
1965	69,811	269	89,910	264	188,488	233	145
1966	62,404	248	107,655	317	219,398	270	204
1967	50,349	193	115,802	341	332,649	411	255
1968	50,164	192	165,787	485	450,166	558	313

One factor in this reversal was the universities' structural advantage in the budgetary process. This helps explain the overall strength of the universities, but not the sudden rise in 1967 and 1968. This rise can be attributed to pressure from students and potential students (especially from those denied admission to universities as a result of the entrance examination), their allies in the media, and from individuals well connected to the policy elite.

Entrance to Brazilian universities is controlled by an examination called the *vestibular*. The number of actual places available in the universities, usually about one-third the number of candidates, is known before the examination is taken. But there is a category of *aprovados,* or approved candidates, composed of those scoring above a certain minimum level set as a standard of performance on the test. Generally, there are about twice

43. This emphasis on the formal budget does not deny the importance of the cash budget, which involves the Ministry of the Treasury much more significantly, but informants suggested that in education the advantages gained at the early stages of the budget process carried over into final expenditures. On the importance of the cash budget, see Robert Daland, "The Paradox of Planning," in Rosenbaum and Tyler, *Contemporary Brazil,* p. 42.
44. *Anuários* as cited in n. 40.

as many approved candidates as places. The approved but unplaced candidates are the *excedentes*. The potential influence of the *excedentes* might
appear slight since they were unorganized and lacked the locational advantage of actually being in a university, but between 1967 and 1969 they
received extensive and sympathetic coverage in all the newspapers. Moreover, they were explicitly seen as a potential threat to social peace by some
interviewed policy-makers.

The *excedentes'* problem was compounded in 1967 when the government
was attempting to reduce budget deficits by cutting expenditures in all
ministries. The number of places offered to university students was *reduced*
between 1966 and 1967. When the number of *excedentes* became known
in 1967 and 1968, numerous editorials and magazine articles demanded
government action. Student demonstrations articulated the demand for
more *vagas* (places) in the universities. And informants reported considerable pressure on the president from the rectors and from nongovern-
ment influentials who had access to key policy-makers. Such pressure is not
surprising, because in a university system as inadequate as Brazil's, even
people with relatively high socioeconomic status may be disadvantaged.
Pressure to expand enrollment is likely to come from people with considerable political muscle—something which is clearly not the case for
primary education.

Student political activity increased sharply in Rio and São Paulo between April and August 1968.[45] A dispute over a student restaurant widened
after police killed a student in the streets of Rio. In this crisis the Minister
of Education, Tarso Dutra, was removed from authority by being sent to
Porto Alegre while military personnel directed negotiations with the students.

Though the adequacy of federal funding was among the issues of the
strike, the degree to which the media associated student violence with the
lack of funds is nevertheless surprising. That *Correio da Manhã,* normally
somewhat leftist, would take this position was expected; but it was unprecedented for the Information Bulletin of the São Paulo Stock Exchange
to echo such a sentiment.[46] The government's response indicated that it,
too, was aware of the linkage. In 1967 all the *excedentes* were placed in
some university. In 1968 over one hundred thousand new spaces were
created, many in universities opened that year, and the budget included a
huge increase at the university level.

The response of the federal government was not caused by the technical

45. Student political activity was not absent between 1964 and 1967. But it was not
until the latter date that the *excedentes* problem became so serious that the
media supported funding demands in student protests. See Robert O. Myhr,
"Student Activism in Development," Rosenbaum and Tyler, *Contemporary
Brazil,* pp. 349–70.
46. *Revista dos Mercados,* 19: 213 (May 1968), p. 1.

studies of its educational planners. Not only were they advocating increases in secondary education, but they were actually predicting unemployment or underemployment among professionals if the university system continued to expand.[47] The government's response stemmed instead from the pressure of students without university places.

These decisions underline the importance of nontechnical inputs in the policy process. Sharp increases occurred in the whole educational budget in 1965, but these antedated real planning. 1967 and 1968 showed large increases in university spending, but they clearly resulted from student— not *técnico*—pressure. *Técnicos* desired large increases, but in the absence of allies they seemed unable to prevail.

As a result of their weakness in the policy process, the *técnicos* in IPEA and in the Human Resources Center had low opinions of their own efficacy. Center staff did not believe "technical criteria originating in IPEA" were important in deciding the size of the education portion of the federal budget or its sectoral division. They often mentioned "political pressure" as a powerful determinant, and they noted that their own influence was less than that of the rectors and friends of the minister.

These feelings of inefficacy may have resulted in a tendency for the studies and reports of IPEA to focus on particularly narrow projects, while in conversations *técnicos* spontaneously expressed concern about broader issues. Some respondents specifically believed that they could affect policy only if they worked on narrow low-cost projects. Such a narrowing might have favorable consequences for the *técnicos* but negative consequences for the broader policies they espoused.

Técnicos' inability to influence major policy questions may also have led to a withdrawal from potential conflict situations. When IPEA staff members were asked what they had done to promote their goals, they often answered that they had not attempted to promote policy goals beyond sending their reports to the appropriate recipient. Clearly the potential strategies available to bureaucrats, including encouraging clienteles, mobilizing support, and searching for allies, were much more limited than in an open polity. But their extreme quiescence may have been a rational response to their negative experiences in the policy process.

47. Albert Fishlow has pointed out that Brazilian *técnicos* seem to go through periods of optimism and periods of pessimism. My interviews were carried out at a time when pessimism was shifting to the current (1970–1972) euphoria. Unemployment of professionals really depends on the general state of the economy. In 1969, however, there were three thousand doctors in Rio not practicing medicine. In the absence of coercion or the kind of altruism that would encourage medical practice in the countryside, producing more doctors may accentuate urban unemployment or underemployment. This is also true of engineers, economists, and other professionals.

Conclusion

The Policy-Making Process

Formal governmental positions were more likely to be filled by *técnicos* after the coup. Both in education and housing, increased centralization and shifts in authority to certain ministries increased the percentage of *técnicos* in formal authority positions. In salary policy, pre-1964 decisions were also made by *técnicos,* but the ideological unity and loyalty of decision-makers increased in the post-coup period.

Evidence for differences in decision-making behavior among different kinds of personnel was inconclusive. Differences emerged between the behavior of the groups making salary and *favela* policy. Members of the *favela* group had a much less evidence-based decision style, and their emphasis on the necessity for "global planning" seemed to rationalize inaction.[48] But while the *favela* decision group included both *técnicos* and non-*técnicos,* no consistent differences in behavior were found between the two. This suggests that the nature of the groups and the kinds of problems they must solve are important determinants of decisional behavior.

The *abono* group members had high positive affect, faced a specific task, and would be negatively evaluated by their ministers if they failed to solve it. The *favela* group participants had a mandate to solve the whole *favela* problem of the State of Guanabara, but they had strong incentives to do nothing, because action invited penalties and would bring few rewards. Decision-making behavior might be quite different for each group in a different setting.

A second aspect of technically oriented behavior is the criteria used to choose among alternatives. Even though participants in the *abono* group were much more likely to evaluate alternatives according to evidence, their criteria were political rather than technical. Their goals included getting the solution accepted by the ministers, enforcing it, and so on. These criteria were unrelated to the effects of the wage increase on living standards, aggregate demand, or inflation. Moreover, the outcome in the *abono* group was clearly affected by shortages of time and deficiencies in data.

In spite of similarities in decision-making behavior between *técnicos* and non-*técnicos,* technically trained personnel might gain influence because the elite listens to them. This seems to have occurred in salary policy, where *técnicos* were, if not supreme, certainly dominant. The other cases, how-

48. Richard Fagen points out that Cuban revolutionary leadership has an anti-planning mentality precisely because of its belief that "too much thought given in advance to the possible consequences of a program tends to erode revolutionary will and courage." See Fagen, "Continuities in Cuban Revolutionary Politics," *Monthly Review,* April 1972, p. 34.

ever, show the limitations of *técnico* influence. While *técnicos* filled formal governmental positions, real influence was retained by people who did not hold official posts. Certain people in *favela* policy, for instance, influenced outcomes even though they represented the kind of particularistic interest supposedly alien to the regime. Advocates of experimental solutions, regardless of technical knowledge, fared poorly because of the paternalistic and anti-lower-class values of high officials. In allocating educational resources, *técnicos* were also mostly ineffectual. They did not generally advocate increased spending on universities; the policy elite responded to the demands of students and university administrators. The elite talked about planning, but such talk was really a screen used to justify decisions made on other criteria.

The Incompleteness of Goals

The coup elevated to power men with goals impacting on some policy areas but not on others. If a policy problem fell into an area affected by regime goals, outcomes would be fairly predictable. But what if no intensely held common goal applied?

Apparently "what happened" was that the regime acted to preserve political and social calm. Both the storm of criticism which arose following the rains of 1966 and 1967 in the State of Guanabara and the pressure from rectors, parents, and university students in 1967 and 1968 came from sources either friendly or neutral to the regime. Neither could be repressed by military action or political persecution. In both cases the regime responded with vigorous measures designed more to alleviate the immediate crisis than to solve the long-range problem. The *favela* problem was federalized, thus meeting the criticism that the State was incompetent. But the regime was unwilling to commit enough resources to deal seriously with squatter settlements. Similarly, the initial response to the pressure for higher university enrollment was to create more places without increased expenditures, producing overcrowding. Later, expenditures did go up sharply, but at the cost of decreases for other educational levels and without consideration for the effects on the labor market.

Since the desire to improve decision-making supposedly applied to all policy areas, it might be expected that, once a new policy problem arose, decision-making in that area would be modified and *técnicos* would gain influence. This did not occur. In *favela* policy Carlos Costa's ability to force *favela* removals ended only upon the death of Costa e Silva; in allocating educational resources, the rectors were far stronger than the *técnicos*. *Técnicos* in education competed with people better connected to the president and ministers. As long as technical inputs themselves were not valued highly, the ability of *técnicos* to achieve their objectives through bargaining was limited, because *técnicos* could operate only inside the government and depended upon an organizational hierarchy. So *técnicos* did little to promote their own policy goals, but it is doubtful that more activity would have yielded greater success.

Toward a Ranking of Elite Goals

The findings of this essay help differentiate between regime rhetoric and real priorities. While economic goals seem to have been foremost both in rhetoric and actual behavior, it proved impossible to develop a consistent ranking of goals beyond them. For example, the elite implemented its desire to end excessive fractionalization of parties, demagogic labor leadership, and the *jogo de política*. But this implementation was much more thorough in some policy arenas than in others. It was effective in economic policy: the repression of labor and centralization of policy-making power in the Ministries of Planning and Treasury facilitated a rigorous wage policy. But in *favela* and education policy, particularistic forces regained power, and *técnico* decision-makers were overruled by regime clienteles. Similarly, the regime's desire to institute rational policy-making was partially realized in economic policy, given limitations of time and data. But *técnicos* were scarcer in *favela* and education policy and received little support even when they participated in the policy making process.

In sum, in those policy arenas in which the regime had intense goals *and* in which technical expertise was traditionally utilized, the regime's desire to rationalize policy-making was at least partly implemented. But in arenas in which the regime had no intensely held goals and in which the *jogo de política* had traditionally held sway, the regime seemed unable or unwilling to modify policy-making. In these latter arenas policy-making was considered rationalized once the regime had attained power. Any outcomes were considered better than pre-coup outcomes simply because the regime had produced them.

PART VI

Europe

MILITARY INTERVENTION
AND THE POLITICS OF GREECE

James Brown

In this century military intervention in the political arena by the Greek armed forces has been the rule and not the exception.[1] The forms of and the motivations for this political activism have varied greatly, but, in spite of the often decisive effects of their intervention in political affairs, the Greek military and the reasons for their intervention have not been studied in a serious or systematic manner.

The most recent intervention by the Greek armed forces (primarily the army) was a bloodless coup d'état which took place on April 21, 1967. The world was suddenly jolted into wondering if Greece had been transformed into the image of an African or Latin American nation, when the rumbling of tanks in the early morning hours was substituted for the popular or parliamentary vote as a normal procedure for changing governmental leadership. Although many analyses have appeared which attempt to explain the collapse of democracy in Greece, very few of these have approached this topic as an object of serious scientific study; and the same can be said for the related topic of why a professional officer corps intervened in the political arena.

1. Seven coups d'état have taken place in this century: 1909, the military group led by Col. Zorbas; 1922, group led by Col. George Plastiras; 1933 and 1936, led by Gen. John Metaxas; 1926 and 1935, led by Gen. George Kondyles; 1925, led by Gen. Theodoros Pangalos. In addition, many attempted coups either failed or never got "off the ground," including some as recently as 1951 and March and early April of 1967.

It is the intent of this article to examine two critical questions relating to the April 21 coup. First, what was the nature of the political, social, and economic conditions within Greece from 1963 to 1967 that might have led to the coup d'état? Second, given the social and ideological characteristics of the Greek officer corps, what was their predisposition toward the political arena?

Although the approach of this analysis is microanalytic, it is hoped that this study will promote a better understanding of comparable problems in other places and different times. This essay seeks to examine some of the broad generalizations of the nature and conditions of military intervention, using the April 21 coup as a "test case." Some scholars might object that this is scarcely legitimate—that just one case or one instance is insufficient for testing purposes. Or the alternate argument might be raised that there are peculiarities in the Greek case which make it unique. In defense of this study one can argue that the test of any generalization is that it should cover every instance from which the generalization is drawn. Second, though one can admit certain special features of the Greek armed forces as a military institution, one can deny that they are in any way "unique." As a matter of fact, all armed forces have special features and characteristics about them, but this is not to say that they are, therefore, necessarily separate and different when regarded as institutional complexes.

Economic, Social, and Political Conditions, 1963–1967

A starting point in examining the economic, social, and political conditions within Greece during the 1963–1967 period is to turn our attention to a study by Chalmers Johnson entitled *Revolutionary Change,* in which the author points out that, in order for a revolution or coup d'état to take place, there must exist some sort of "accelerators that hold forth the promise to the revolutionaries that they can break the elite's monopoly of force." [2] The author seems to imply that there are specific conditions, situations, or events favorable to the implementation of a revolution or coup d'état. In other words, a favorable opportunity or occasion must be present before the revolutionaries take definitive action, which opportunity may be inherent within a situation at a particular point in time. But the crucial questions still remain as to exactly what conditions are more likely to enhance the probability for success.

Examining the economic conditions in Greece that might have heightened the probabilities for the April 1967 military intervention, we will utilize Davies' economic model, which introduces the concept of "need satisfaction." [3] He hypothesizes that revolutions or coups d'état are more likely to

2. Chalmers Johnson, *Revolutionary Change* (Boston: Little, Brown, 1966), p. 152.
3. James Davies, "Toward a Theory of Revolution," *American Sociological Review,* February 1962, pp. 5–19.

come about when a prolonged period of growth in economic or social development is followed by a short-term phase of economic stagnation or decline. The result of this "J-curve," as he calls it, is that the soaring expectations in the minds of the populace created by such economic growth usually go beyond the actual material satisfaction of the needs.[4] Thus, a successful coup d'état or revolution is the work of neither the destitute nor the well-satisfied, but of those whose actual situation in life is improving less rapidly than they expect.

On the other hand, we might advance the alternate hypothesis that when there is a prolonged period of economic decline in the need satisfactions of the populace, a frustrated military may view this as an opportunity or occasion to instigate a coup d'état. Thus arises the necessity of investigating the Greek economic indices that measure economic prosperity or depression and which, in turn, may have led to the 1967 coup: i.e., gross national product, consumer price index, wages and employment figures, balance budgets, and balance of payments.

The economic picture in Greece during the period under investigation, 1963 to 1967, was an uncertain one at best, although in the latter part of 1966 confidence in the economy began to rise again slightly and a marked improvement in public finances made possible a substantial increase in public investment expenditures. For the four years immediately preceding the 1967 coup, the rate of growth in the gross national product averaged eight percent.[5] This was primarily a result of expanding industrial production which increased by more than fifteen percent in total value during 1965–1966, even though the agricultural sector was lagging behind.[6] Between 1960 and 1964 consumer prices remained relatively stable, but in 1965 inflationary pressures began to build up, causing rapid increases in the price of consumer goods. This rise during the course of 1965 and 1966 averaged about six percent.[7] The rise in the consumer price index (see tables 1 and 2) seemed to be mainly the result of measures taken by the government in the direction of increased public expenditures and was especially manifest in the areas of uneconomic agricultural subsidies, rising incomes, and an unfavorable balance of payments. Wages during the years between 1964 and 1967 showed an average increase of approximately 8.4 percent. Although the wage scale did indicate an increase, unemployment remained relatively high, especially in the agricultural sector, on the average about five percent for the same period.[8] So, in essence, from 1963 to 1967 the economic climate could not be said to be favorable, although in 1966 the economy slowly

4. Ibid., p. 6.
5. *An Economic Report on Greece* (London: Lloyds Bank, Ltd., 1967), p. 4.
6. Ibid., p. 8.
7. *OECD Economic Surveys, Greece* (Zurich, Switzerland: 1967 and 1969), pp. 10–12.
8. Ibid., pp. 13–20.

Table 1: *Main Economic Trends, 1964–1968*
Economic Growth and Imbalances

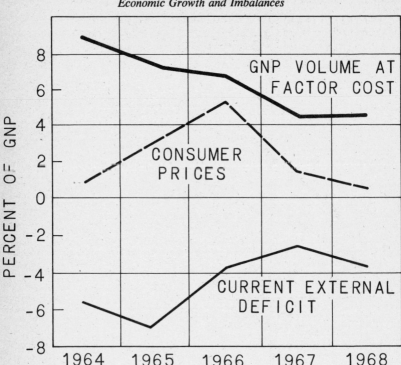

Source: Greek Submission to OECD. *OECD Economic Surveys*, *Greece*, February 1969.

began to right itself. Thus there had been three years of worsening economy, followed by an improvement, precisely opposite to the pattern set by Davies' "J-curve." So at first glance the Davies hypothesis is not satisfied.

The second hypothesis, that is, regarding the prolonged period of economic deterioration that might be interpreted by the military as sufficient opportunity to instigate a coup d'état, is a much closer fit, even though the 1966 economic upturn in Greece is inconsistent. While there is this additional support for the hypothesis of economic decline, another conditioning variable needs to be introduced at this point.

It is not enough merely to examine these hard economic indicators. One must carry an analysis a step further and examine how these economic variables are linked to the masses. In order to do so, we will examine public opinion data for the period of 1964 to 1967. An investigation of such data might provide us with some clues regarding the expectations of the populace,

Table 2: *Price Indices, 1965–1968*
Quarterly Averages

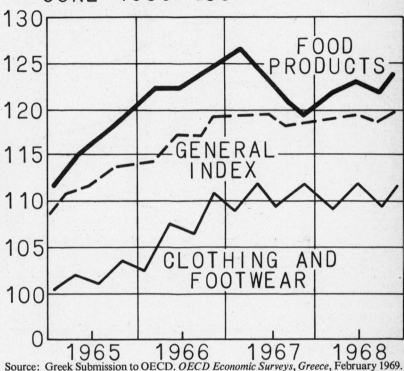

CONSUMER PRICES
JUNE 1959-100

Source: Greek Submission to OECD. *OECD Economic Surveys, Greece*, February 1969.

and in turn how this might have been perceived by the officer corps as a favorable occasion for a coup.

In table 3 it is apparent that those surveyed in the year 1966 were quite negative about their prospects for a higher standard of living. This correlates with the problem of inflation that Greece was suffering during that year. Prior to that time, and subsequently in 1967 and 1968, the outlook of those surveyed was surprisingly optimistic regarding the same standard of living issue. More particularistic data concerning the economic outlook and how it affected individuals is offered in tables 4 and 5. Judging from this data the respondents appear to fall into the pessimistic category for the period of 1966 and 1967, apparently anticipating a year of rising prices, industrial disputes, an increase in taxes and economic difficulties (all this, even though

Table 3: *Greek Populace Outlook Toward Standard of Living*
1964–1968 *

	1964	1965	1966	1967	1968
Going Up	36%	36%	15%	47%	56%
Going Down	30%	18%	67%	17%	12%
Same	33%	40%	13%	29%	25%
Don't Know	1%	6%	5%	7%	7%
Sample Size	400	800	400	400	800

Source: Social Surveys Limited, London, England.
* Text of Question: Speaking generally, would you say that your standard of living
(things you can buy and do) is going up, going down, or remaining
the same?

the employment picture was somewhat improved), all of which substantiates
our earlier economic data.

An interesting fact that emerges from our data is that subsequent to the
coup, in 1968 our respondents appeared to have a much more optimistic

Table 4: *Future Outlook in the Social and Economic Spheres*
1965–1969 *

	1965	1966	1967	1968	1969
A year of rising prices	58%	74%	59%	23%	24%
A year of falling prices	6%	7%	5%	29%	22%
No opinion	36%	14%	36%	48%	54%
A year of full employment	38%	41%	31%	48%	67%
A year of rising unemployment	19%	32%	30%	11%	6%
No opinion	43%	27%	39%	41%	27%
A year of strikes and industrial disputes	19%	39%	37%	4%	3%
A year of industrial peace	31%	28%	21%	72%	85%
No opinion	50%	33%	42%	24%	12%
A year when taxes will rise	29%	84%	54%	14%	18%
A year when taxes will fall	21%	4%	8%	32%	23%
No opinion	50%	12%	38%	54%	59%
A year of economic prosperity	25%	11%	14%	50%	65%
A year of economic difficulty	33%	68%	46%	12%	9%
No opinion	42%	21%	40%	38%	26%

Source: Social Surveys Limited, London, England.
* Text of Question: Which of these do you think is likely to be true for (year)?
These surveys were conducted in December of the previous year.
With the exception of 1965, the sample size was 400; for 1965 it
was 800.

Table 5: *Future Outlook Toward a Better or Worse Year**

	1964	1965	1966	1967	1968	1969
Better	59%	25%	23%	62%	60%	67%
Worse	16%	37%	40%	6%	6%	3%
Same	8%	17%	7%	12%	13%	26%
Don't Know	17%	21%	30%	20%	21%	4%
Sample Size	400	800	400	400	400	400

Source: Social Surveys Limited, London, England.

* Text of Question: So far as you are concerned, do you think that (year) will be better or worse than (year)?

outlook, in spite of the fact that the economic picture was far from favorable. We might speculate that so soon after the coup our respondents were extremely reluctant to criticize the junta and its governing capabilities and thus were reluctant to express anything but a positive viewpoint.

Substantive information about the linkage between economic disequilibrium and a coup d'état is still somewhat remote, but we can arrive at some tentative conclusions regarding economic data vis-à-vis the attitude of the populace about their economic future. As Martin Needler points out, "The overthrow of a government is more likely to occur when economic conditions are deteriorating." [9] This, however, was not true in this case. Tensions and frustrations were evident from our data and we might infer that apprehension was the prevailing mood of the Greek citizenry. Although our economic data indicates a direction of economic stability in 1966, this positive effect is not translated in our public opinion data. A gap seems to exist between the economic indicators and their correlating effect upon the man in the street. In 1966 and 1967 a time-lag effect was apparent in that the populace was still economically depressed. One important conclusion for comparative civil-military relations is the gap between objective economic statistics and the subjective perceptions of the population. Great care should be utilized in studying economic data as a means of testing civil-military propositions. The economic causes of the "opportunity or occasion" for intervention are generally related to the satisfaction of the general public, not the actual health of the economy. They may or may not be related, and as this case indicates, there may be a significant time-lag in the accuracy of popular attitudes. Analysts should, therefore, remember that hard data are at best indirect indices of the attitudes that are conceptually more closely related to opportunity.

There is a further causal step required for the link between deteriorating

9. Martin C. Needler, "Political Development and Military Intervention in Latin America," *American Political Science Review,* September 1966, p. 624.

economic conditions, or, more accurately, popular perceptions of deteriorating economic conditions, and actual factors leading to military intervention.
If it is a factor creating the opportunity or occasion, then the coup leaders
must perceive it as such. So surely, if the state of the Greek economy had
been a factor seen as creating an opportunity for intervention, the colonels
would certainly have used it in their rationale and stated public positions.
This, however, was not the case. In none of their post-coup pronouncements
or in the interviews that this writer conducted with the Deputy Prime Ministers, S. Pattakos and N. Makarezos, and with various military officers, were
the economic conditions in Greece from 1963 to 1967 ever mentioned as
reason for the coup d'état.[10]

The Greek officer corps, of course, knew of the economic data discussed
above, and would have seen the upturn, but other figures must be taken into
account if this be the case. In a revolutionary or "coup" situation, which is
fundamentally concerned with a rather small group of politically active individuals, the issues of budget and balance of trade may greatly affect elite
mood. Perhaps this is because government deficits and a decline in the
balance of trade so gravely affect the opportunities for career advancement
and overall prosperity of the military. Thus, although these two indicators
may not directly affect the general public initially, they are closely related

Table 6: *Imports, Exports, Balances of Trade, and Balances of Budget*
(In Millions of Drachmai)

Year	Balance of Budget	Exports	Imports	Balance of Trade
1964	+1.2	9,256	26,556	−17,296
1965	−0.4	9,833	34,012	−24,179
1966	+2.1	12,179	36,685	−24,506
1967	+0.3	14,856	35,588	−20,732

Source: *Statistical Yearbook of Greece* (Athens, Greece: National Statistical Service,
1969).

to the perceptions of would-be conspirators and dissident elites, who are the
prime movers of a coup d'état. In other words, these two indicators may
serve as the opportunity or occasion, if not the motive, for intervention. By
utilizing the balance of trade and budget figures as economic indices in relation to the frame of mind of highly politicized individuals, we might expect
to find that the trend in the balance of trade and balance of budget would
drop sharply prior to the April 1967 coup.

10. Interviews were conducted with the deputy prime ministers in November
    1971.

We should expect to find some correlation between the decline in the balances of trade and the balances of budget and military intervention. Such was not the case. A partial explanation might be that the Greek armed forces were basically dependent on the United States to supply them with the latest arms and equipment. The estimated value of the anticipated military arms for 1967 was $64,981,000.[11] So it would seem that any imbalance in trade and budgets certainly did not threaten to disturb immediately the normal functioning and operation of the military.

As we have indicated earlier, inflation was a disturbing and vexing problem during the 1963–1967 period. It was not, however, severe to the point of creating economic failure. The balance of payments picture was surely negative in that imports at that time were two and a half times as high as exports, but the overall economic picture was sufficiently stable so as to rule out the state of the economy as a precipitant of the coup d'état. In fact, at no time during this writer's interviews with Greek officials did they indicate concern about the economic status or mention that they had been affected in any way by the economic conditions of the 1963 to 1967 period.

In summary, we must conclude that the economic evidence presented does not allow us to conclude that the economic factors were involved either in the opportunity or occasion for intervention. The Davies model is directly disconfirmed in our analysis of the April 21 coup, while the theory of general economic deterioration has positive support when its subjective form of popular support is considered. But before weighing the overall importance of economic factors, we must consider the political and social factors said to be important determinants for military intervention.

The prevalence of riots and strikes may affect the legitimacy of parliamentary government. Strikes were prevalent to the point of being rampant in Greece from 1965 to April of 1967 and, in fact, were anticipated by the populace (see table 4). During this period most of the trade unions went on strike at various times: an estimated 950 strikes took place, or about 24 per month.[12] It is further estimated that of the above-numbered strikes a majority were either organized or inspired by Communist or Communist-front organizations. The principal organizations involved in these strikes were the General Confederation for Labor, the Lambrakis Movement, the United Democratic Left (EDA), and some members and followers of the Center Union Party (EK). All of these elements were more than willing to embark upon an extraparliamentary course of action. The Lambrakists, in particular, were extremely active in the demonstrations and other assorted antigovernmental activities. In fact, summarily, all of these organizations had

11. Letter from Department of State, Agency for International Development, 18 September 1967.
12. *Diati Eginai H Epanastasi tis 21 Apriliou, 1967* (Athenia: Government Printing Office, 1968), p. 95. Also see the *New York Times* for the period of 1966 to April 1967.

expressed an intolerance, at once, of the traditional right-wing politicians, parliamentary institutions, and the armed forces.

Not only were strikes prevalent, but during this same period rioting seemed to be the order of the day, resulting in many injuries and some losses of life. From 1965 to April of 1967 approximately 1,200 individuals were injured in such riots and 15 lost their lives. Of those injured, 300 were gendarmes.[13] Since the government of Greece is so intimately involved in its economy, strikes and disruptions of any kind can be completely misinterpreted. Threats of military call-ups were rife and on several occasions the army was alerted to quell the rioting.[14] But the bulk of responsibility for putting down these disorders was left to the local gendarmerie, who were known far and wide for dealing with such disturbances with a heavy and none too discriminating hand.

In connection with the above societal conditions, the relative stability or instability of the parliamentary government was another central issue of the times. Six separate governments came to power during this time, averaging a new government about every five months.[15] This instability, in itself, might have undermined the legitimacy of these governments, but their ineffectiveness was further compounded by the behavior of the members of Parliament, who seemed unwilling to work within the framework of the institutional rules and procedures, and who must, therefore, bear some of the responsibility for the breakdown of democracy in Greece. Numerous accounts in the press reported that members of Parliament actually threatened other fellow members with bodily harm and even scuffled and engaged in "fisticuffs" on occasion.[16] Such situations, although not directly linked to the question of governmental legitimacy, assuredly raise considerable doubt as to the institutional viability of Parliament as a body, as well as to the stability, both politically and mentally, of its members. We should hasten to point out that Parliament might well have reflected external tensions and still have remained viable if the conditions had been different.

At no time during this chaotic period was martial law imposed, nor were the rights of assembly or petition denied or banned. As mentioned earlier, civilian authorities did not seem overly dependent on the armed forces for the maintenance of peace and order, but the above configuration does indicate that the various governments were unable to resolve the myriad conflicts through compromise and bargaining.

13. *Diati Eginai H Epanastasi,* p. 81.
14. Interview with a knowledgeable officer of the Greek General Staff, Athens, June 1969.
15. The six governments (or caretaker governments) were: George Athanasiades-Novas, 15 July 1965; Elias Tsirimokas, 24 August 1965; Stephanos Stephanopoulos, 25 September 1965 and 15 July 1966; Ioannis Paraskevopoulos, 14 January 1967; Panayiotis Kanellopoulos, 15 March 1967.
16. *Diati Eginai H Epanastasi,* pp. 48–49. Also see *Parliamentary Session Papers* for 1965–1966.

In contrast to economic factors, there have been frequent public announcements made by the junta to justify their coup in which they continually mention the social and political turmoil that existed during the period from 1965 to April of 1967 as the reason for the coup. The instability qua instability of this period seems to have served as the opportunity for intervention. That is to say, the prevailing social and political conditions at that time provided the conspirators with what they took to be a favorable chance to intervene. Ironically, however, according to a poll of the populace during 1965–1966, the majority (forty-three percent) felt that the most crucial problem then facing Greece was economic in nature, while twenty-seven percent of those polled felt that the political problems were more acute.[17] Thus, in actuality, the negative economic expectations of the population were an important, if not the most important, factor of the opportunity, but it was not perceived as such by the junta leaders. The success of the coup might have been shaped by economic factors, but the perceived opportunity or occasion must be viewed through the eyes of the officer corps. Therefore, our findings suggest that political instability is a telling variable in signaling to the military that intervention is possible of success.

Political Perspectives of the Officer Corps

Although the previously discussed societal conditions exist, individually or collectively, in a given political system, it does not necessarily follow that a military will consider intervention even if the likelihood of success is on the high side. According to Huntington, professionalism is a decisive factor in keeping the soldier out of politics.[18] However, professionalism, in and of itself, will not keep the military from intervening in politics. Finer indicates that military professionalism often thrusts the military into a collision course with the civilian authorities.[19] To inhibit this tendency the military must absorb the "principles of the supremacy of the civil power," [20] which we can define as the acceptance among military men, both formally and effectively, of the major policies and programs of the government— which policies, furthermore, "should be decided by the nation's politically responsible civilian leadership." [21] In other words, seizure of control by the military normally will not occur when the officer corps recognizes civilian control of the armed forces as legitimate.

17. *Elpides Kai Phovie Dia tin Ellada, 1966* (Athens: Institute of Anthropos, 1967), pp. 354–55.
18. Samuel P. Huntington, *The Soldier and the State* (New York: Vintage Books, 1964), pp. 8–18.
19. S. E. Finer, *The Man on Horseback* (New York: Frederick A. Praeger, 1962), pp. 25–27.
20. Ibid., p. 28.
21. Burton Sapin and Richard Snyder, *The Role of the Military in American Foreign Policy* (New York: Doubleday Co., 1954), p. 52.

Another researcher, M. D. Feld, suggests that social and political factors, not military professionalism, elucidate modern civil-military relationships. He believes that political stability is a prerequisite for, and not a consequence of, the creation of a professional, "apolitical" officer corps.[22]

As far as Greece is concerned, the hypotheses of Finer and Feld seem to have greater validity than Huntington's. The Greek officer corps is a thoroughly professional organization, according to the criteria of professionalism established by Huntington.[23] The Greek officer corps professes to be "above politics." To be "above politics," according to Janowitz, requires one to be nonpartisan on political issues.[24] However, being above politics in the Greek officer corps does not carry with it Janowitz's dictum of being "apolitical." As a result, whenever a challenge has been raised as to the legitimacy of civilian control of the military, or political institutions have been unable to mediate between various groups within Greek society, the military has felt justified in intervening in order to correct or redirect the political arena. Greek officers interviewed made a very fine distinction between the state or the "national interest" and the civilian governments in power.[25] For example, one officer stated: "When the politicians failed to solve the political crises during the 1966–1967 period, it was necessary for the armed forces to save the country from this political chaos. It was an 'Epanastasis dikea' (a just revolution)." Another stated that "Greece had natural enemies from within and on her borders; thus the Army must always be vigilant. They are the protectors and saviors of Greece." Professor D. Tsakonas stated that "the role of the officer corps in the Balkans differs from his counterpart in the United States and in Western Europe. The officer corps in the Balkans is also a policeman. That is to say, they must protect themselves; they have no choice." And lastly, Deputy Prime Minister S. Pattakos stated to this writer that "we (the officers) could not allow the chaotic political situation to continue. Therefore, we told the King and the General Staff that some corrective action had to be taken. When nothing was forthcoming, we had no alternative but to make the Revolution and save the country. And we did so without bloodshed."

Historically, the political allegiance of the officer corps had usually been dictated by alliances with the monarchy, politicians, and political parties. From 1912 on the armed forces were split into two rival groups: the Republicans and the Royalists. In the post-World War II era the prime ad-

---

22. M. D. Feld, "Professionalism, Nationalism, and the Alienation of Military," in Jacques Van Doorn, ed., *Armed Forces in Society* (The Hague: Mouton, 1968), p. 68.

23. Huntington, *Soldier and State*, pp. 11–18.

24. Morris Janowitz, *The Professional Soldier* (New York: The Free Press, 1960), p. viii.

25. Interviews were conducted with Greek officers during the spring of 1969 and the fall of 1971.

versary of the military was the Left. This was principally a result of the 1946–1949 Greek Civil War—the bloodiest fratricide in the history of modern Greece. The Civil War left an indelible impression on the officer corps and marked a new era for them—a period in which the armed forces emerged as a major power bloc independent of traditional parliamentary factions and leaders. The military was now viewed by the populace and by its own members as the symbol of national unity that had saved the nation from a Communist takeover.

It might be appropriate to compare the Greek military at this juncture with the armies formed by national liberation movements elsewhere. The military now possessed a combination of pragmatic and ideological commitment, a cohesion formed by common experiences and a heightened sense of self-esteem.[26] Now the armed forces, and especially the army, felt that their main purpose was to protect the nation from Communism, both from within and from without, an ideological position of hard-line anti-Communism. Thus, the military viewed the Left as its major opponent, and any action taken by the politicians that might be interpreted by the military as concessions to the Left were difficult for the armed forces to bear and could conceivably have led to a coup d'état.[27] The Left, on the other hand, has viewed the armed forces and the monarchy as the primary compromisers of Greece's independence.

During and immediately following the Civil War, draftees and left-wing officers were separated from the others and sent to a small island off the tip of Attica for military training and indoctrination. Upon completion of this training the draftees were put into special units that were widely known for their severity, and they were usually assigned to the most unpleasant duties, such as mule driving in the Rhodope Mountains. It is a fundamental and firmly adhered to assumption of the professional officer corps today (and, to some extent, prior to the coup) that no person with leftist political ties could possibly be loyal to Greece.[28]

For its part the Left has continually accused the military of fascist tendencies. The process of military indoctrination and the backgrounds of high-ranking officers have been criticized in like manner. Left-wing proposals for the normalization of political life have included such items as amnesty for the veterans of the Civil War on the guerrilla side. The military, and even some moderate elements in Greek society, view these individuals as traitors, but the EDA lauds them as the real exponents of Greek nationalism and independence. The EDA naturally opposes the NATO alliance and all

---

26. M. Janowitz, *The Military in the Political Development of New Nations* (Chicago: University of Chicago Press, 1964), pp. 47–48.
27. Discussions and serious planning had taken place in the Greek officer corps for coups d'état in 1951 and early in 1967.
28. Interviews with officers of the Greek General Staff, Athens, Greece, Spring 1969.

230 — Soldiers in Politics

other ties with the West. As an organization it has advocated decreasing military expenditures on the basis that Greece's neighbors, with the exception of Turkey, are all peace-loving socialist states, and has urged diverting these funds to needed social welfare and reform. Such measures and attitudes are seen by the officer corps as a direct threat to its very existence and are viewed with much alarm by that body. Moreover, many Greek officers interviewed by this writer related many atrocities and losses of life that occurred during the Civil War. In fact, the Communists directly threatened, and in some instances killed, the families of some officers simply because the officer was serving in the armed forces.

Upon examining the biographical data of the officers who instigated the coup and those of comparable rank, one finds that during the Civil War they were serving as lieutenants or captains and were in the midst of the fray in the battle against the Communists. This experience has produced in most of them a psychological condition which might be described as a "saviour complex," and which has had a critical bearing on all their subsequent actions.[29] The officers believe that the armed forces are the embodiment of national ideals, aspirations and dreams, and that they are a fixed and integral part of Greek history. The armed forces, modern Greece, and her history have become inseparable in the eyes of the officer corps. The officers view their professional role from a larger perspective, derived from geopolitical, strategic, and historical considerations. This view was substantiated by interviews conducted with American military advisers who indicated that the Greek officer's social responsibility was vested in the state for "historical reasons," and that the military establishment "did not particularly trust or have strong feelings toward civilian authority."[30] Moreover, such developments as Russia's invasion of Czechoslovakia, the tensions between Israel and the Arab world, the presence of the Russian fleet in the eastern Mediterranean, and student unrest in Turkey were cited by officers as evidence of the importance of Greece and her armed forces to the southern flank of NATO and the West. In fact, the Greek officers felt that their role in NATO was far more important and much more tangible than that of some other NATO nations, i.e., Turkey and Italy.

The threat of Communism, real or potential, was and still is today perceived as salient by the officer corps. As is well known, the perception of a threat is as potent a factor contributing to a given behavior as the actual existence of such a threat.

29. Two events set the stage for the April 1967 coup: first, the discovery in May, 1965 of a left-wing secret organization known as ASPIDA (Shield) and connected with Andreas Papandreou; second, the elections scheduled for May of 1967, which George Papandreou's Center Union and the Left were certain to win.
30. Interviews with American Military Advisors, JUSMAGG, Athens, Spring 1969.

Social Origin of the Armed Forces and the Coup Makers

Professor Janowitz, in his analysis of the American military, writes: "The analysis of the social origins of the military is a powerful key to understanding of its political logic, although no elite behaves simply on the basis of its social origin. . . ." [31]

From comparative, historical, and theoretical perspectives there is every reason to believe that the Greek professional officer corps is heavily recruited from the rural areas—agricultural communities and small towns— and in this respect Greece is not unlike other Western countries such as France and the United States. As Janowitz points out, "The out-of-doors existence, the concern with nature, sports, and weapons, which is part of the rural culture, have a direct carry-over to the requirements of the pre-technological military establishment." [32] In the final analysis the link between rural social structure and military organization is based on the central issue of career opportunities.

The military profession, and in particular the officer corps, serves in Greece as a channel of social mobility wherever there is a relative absence of economic opportunity. Data from table 7 indicates that the Greek officer corps has roots in the rural areas.

Table 7: *Demographic Origins of Greek Professional Officer Corps 1916–1965*

Population Group	Army**	Total Population
Less than 1,000	32.1%	34.6%
1,000–5,000	25.9%	20.2%
5,000–10,000	6.2%	3.2%
10,000–50,000	18.0%	12.0%
50,000 or More*	4.2%	3.5%
Thessaloniki and Athens	13.6%	26.5%

Source: *Statistical Data of the Social Characteristics of the Professional Officer of the Army* (Athens: Armed Forces Headquarters, The Statistical Section, June 1969; writer's translation), pp. 4–5.

* This does not include the areas of Thessaloniki and Athens.
** The Army officer corps consists of about 65 to 70 percent of the total officer corps.

The two major cities in Greece, Athens and Thessaloniki, do not contribute substantially to the officer corps of the Army. Recruitment patterns of the military academies indicate that Athens supplies 39.3 percent of the

31. Janowitz, *The Professional Soldier*, p. 81.
32. Ibid., p. 85.

naval cadets, 25.1 percent of the air force cadets, and only 12.9 percent of the army cadets (see table 10). A partial explanation for the overrepresentation of naval and air force cadets may be the recruitment patterns from different socioeconomic strata. We could argue that the concentration of rural backgrounds of army officers merely represents the national population of Greece. However, the original conclusion continues to hold true when we compare the military leadership with other elite groups in Greek society. In contrast to 64.2 percent of the officer corps whose social origins are in areas of less than ten thousand population, only 26 percent of the industrialists and 38 percent of the professional men are derived from these same areas. Recent studies indicate that recruitment of officers still heavily favors the rural areas, thus continuing to reflect what is outlined in table 8. It also points up the emphasis placed by the governments since April 1967 on recruitment from these areas. In fact, for the first time an attempt is being made to recruit naval cadets from the rural areas, whereas heretofore the navy tended to represent the more affluent sectors of Greek society.

Table 8: *Demographic Origins of Elite Groups*

Population Group	Army (a)	Industrialist (b)	Professional Men (b)	Total Population
Less than 1,000	32.1%	10%	20%	34.6%
1,000–5,000	25.9%	7%	10%	20.2%
5,000–10,000	6.2%	9%	8%	3.2%
10,000–50,000	18.0%	22%	13%	12.0%
50,000 or More	17.8%	30%	31%	30.0%
Born in foreign countries	—	22%	18%	—

Sources: (a) *Statistical Data of the Social Characteristics of the Professional Officer of the Army* (Athens: Armed Forces Headquarters, The Statistical Service, June 1969, pp. 4–5. (b) Alec P. Alexander, *Greek Industrialist* (Athens: Center of Planning and Economic Research, 1964).

To take the above information one step further, table 9 is an examination of whether or not there is any regional concentration in the backgrounds of the officer corps.

From the preceding data it is clear that the army and the air force tend to draw recruits from the same areas, whereas the navy tends to draw more heavily from Athens and the island areas. It is also obvious that the army and air force officer corps are derived primarily from three areas (see table 9)—Peloponnesos, Central Greece, and Crete—which contribute 52.3 percent and 59 percent, respectively, although representing only 22 percent of the total population. All three of these areas are relatively poor economic

Table 9: *Geographic Origins of the Greek Officer Corps*
1916–1971

Area	Army (a)	Navy (b)	Air Force (c)	Total Population (a)
Peloponnesos	26.1%	16.0%	27.0%	13.1%
Central Greece	15.8%	59.7%	24.5%	3.1%
Attica (includes Athens)	12.8%		16.0%	24.5%
Crete	9.4%	16.1% (d)	7.5%	5.8%
West & Central Macedonia (includes Thessaloniki)	9.2%	5.3%	9.6%	16.5%
Thessaly	7.2%	2.1%	7.0%	8.3%
East Macedonia & Thrace	6.4%	—	1.6%	10.3%
Epirus	6.0%	—	2.0%	4.2%
Aegean Islands	3.8%	—	4.8%	5.7%
Ionian Islands	3.3%	—		2.5%

Source: (a) *Statistical Data of the Social Characteristics of the Professional Officer of the Army* (Athens: Armed Forces Headquarters, Statistical Service, June 1969), p. 8. (b) Random sample of 149 officers, October 1971. (c) Survey of 2,000 officers, conducted in October 1971. (d) Includes all the islands.

regions, and the mobility, security, and prestige offered by the military appeal to many of the inhabitants. The regions of Thrace, Epirus, and Macedonia also contribute a high proportion of the army and air force officer corps. Again, social mobility is a factor, but in addition, these regions border neighboring countries long hostile to Greece; so war, with its attendant chaos

Table 10: *Geographic Origins of Military Academy Cadets Class of 1974*

Area	Army	Navy	Air Force
Peloponnesos	13.8%	8.0%	17.4%
Central Greece	22.9%	30.6%	26.2%
Macedonia	18.3%	4.0%	12.5%
Thessaly	9.5%	2.6%	7.1%
Thrace	5.2%	1.3%	1.6%
Epirus	6.3%	2.0%	2.7%
Crete	5.9%	3.3%	4.3%
Ionian Islands	1.2%	3.3%	0.5%
Aegean Islands	3.3%	4.6%	2.1%
Cyprus	0.1%	0.66%	0.0%
The Capitol, Athens	12.9%	39.3%	25.1%

Source: *Meleth Thn Sxolis Ton Evelpidon, 1971* ["Study of the Army Academy, 1971"] (Athens, Army Statistical Service, 1971).

234 Soldiers in Politics

and devastation, arming, and the use of weapons, may have become ways of
life for the people in these regions. This, then, furnishes ample and active reason for young men to eye a military career.

Since the officer corps is recruited from the rural areas, we may also infer
that they tend to come from rather humble circumstances, and, if so, a
career as an officer does serve as a vehicle for social mobility otherwise unavailable to them. One's father's occupation is utilized as a criterion to measure social origin and may also be employed as an indicator for locating an
officer's parents at some point on the social pyramid.

Table 11: *Officers' Fathers' Occupations*

Occupations	Army (a)	Air Force (b)	Navy (c)	Greek Populace (a)
Farmers	38.9%	38.0%	14.8%	49.2%
Craftsmen	8.6%	4.5%	13.4%	27.2%
Bureaucrats & Tradesmen	23.7%	18.5%	30.2%	18.1%
Professionals	17.8%	16.5%	15.5%	4.5%
Military	11.0%	4.0%	23.9%*	1.0%
Laborers	—	5.0%	—	
Others	—	4.5%	2.2%	

Source: (a) *Statistical Data of the Professional Officer of the Army* (Athens: The Statistical Service, June 1969), p. 9. (b) Survey conducted consisting of 2,000 officers; October
1971. (c) *Meleth Thn Sxoli Ton Evelpidon, 1971* ["Study of the Army Academy, 1971"]
(Athens: The Statistical Service, 1972).
* This figure includes both naval and military careers.

According to the data in table 11, farmers and bureaucrats-tradesmen produce the largest percentage of all officers in the Greek armed forces, which
is just another indication of the aforementioned social mobility. It would
appear that the navy has the highest social base of recruitment, which also
coincides with our previous statements. The naval officers tend to come from
the older, wealthier families.

Another aspect of social origin is the proportion of military officers who
have entered the profession through self-recruitment, that is, who are the
sons of professional officers. The percentage of self-recruitment in the navy
is 23.9 percent, while the extent of self-recruitment for the army is 11.0 percent and 4.0 percent for the air force. Historically, the navy has been the
most prestigious of the services, with links to the royal family and to many
of the affluent shipping magnates of Greece. The air force's relatively small
percentage of self-recruitment may reflect the fact that it is a fairly new

service, as compared to the army and navy, and has not had the advantage
of time to develop self-recruitment patterns.

Janowitz's study of fifty-three nations that have suffered some form of
military intervention disclosed that only two nations—Israel and Pakistan—
have officer corps from predominantly aristocratic families.[33] Evidence from
Janowitz's family background data suggests that most officers from the
nations he studied were essentially denied an active participation in their
country's social and economic order. Can such denial and the possible
resultant frustrations conceivably draw a professional officer corps into
political involvement?

Before we attempt to answer this question, let us examine closely the
social origins of the officers who were direct instigators of the 1967 coup-
d'état. It is noteworthy that, of the seventeen men who were instrumental in
planning and finally executing the coup, four graduated with the class of 1940,
while eight were members of the military academy class of 1943 (see table
12).[34] Of significance is the fact that most of these officers were of field-
grade rank (perhaps the most frustrating rank in any military), from which
level promotions are more difficult to come by and at which point retirement
is usually around the corner. In addition, these classes of young officers from
1941 onward experienced the German occupation of World War II, during
which time many of them participated as members of the resistance in Crete
and the Middle East, and later went on to fight in the Greek Civil War.

There are many striking similarities between the Egyptian Free Officers[35]
and the Greek Colonels. The Free Officers group were from humble origins
(peasant farmers, small landowners, minor officials) and from all parts of
Egypt, analogous to the Greek officers who spearheaded the 1967 coup.[36]
Colonel I. Ladas reflected this humble origin when he said, "We were all
so poor that we called Papadopoulos the rich man because his father was a
schoolteacher." [37] Another important note regarding the colonels was that
they reached the impressionable formative years of their lives and entered
the academy during I. Metaxes' dictatorship. This was the regime prior to
World War II that placed heavy emphasis on law and order and a rigid anti-

33. Janowitz, *The Military in the Political Development of New Nations*, p. 50.
   In addition, this observation has been further substantiated by John J. Johnson,
   The Military and Society in Latin America (Stanford: Stanford University
   Press, 1964); and Edwin Lieuwen, *Arms and Politics in Latin America* (New
   York: Frederick A. Praeger, 1965).
34. Studies conducted by this writer indicate that there is no favoritism accorded
   either academy class with regard to promotions or favors since the 1967 coup
   d'état.
35. P. J. Vatikiotis, *The Egyptian Army in Politics: Pattern for New Nations?*
   (Bloomington: University of Indiana Press, 1961), p. 45.
36. Ibid., p. 46.
37. *New York Times,* 18 April 1969, p. 33.

Table 12: *Leaders of the 1967 Coup d'état: A Sociological Portrait*

Name	Military Academy Class	April 1967 Rank	Branch	Fathers' Occupations	Where Born
Zoitakis, G.	1933	Lt. General	Infantry	Farmer	Nafpactos
Pattakos, S.	1937	Brigadier	Cavalry	Farmer	Aghia Paraskevi, Crete
Hadjipetros, A.	1938	Brigadier	Artillery	Military	Athens
Karydas, K.	1939	Colonel	Armored	Unknown	Patras
Kotselis, P.	1940	Colonel	Infantry	Unknown	Argos
Ladas, I.	1940	Colonel	Infantry	Farmer	Dyrakhion, Megaloupolis
Makarezos, N.	1940	Colonel	Artillery	Farmer	Gravia
Papadopoulos, G.	1940	Colonel	Artillery	Schoolmaster	Eleochorion
Aslanidis, C.	1943	Lt. Colonel	Infantry	Farmer	Halkidikis
Balopoulos, M.	1943	Lt. Colonel	Artillery	Unknown	Athens
Ioannides, D.	1943	Lt. Colonel	Infantry	Farmer	Athens
Lekkas, A.	1943	Lt. Colonel	Infantry	Unknown	Piraeus
Mexis, A.	1943	Lt. Colonel	Infantry	Unknown	Poros Island
Papadopoulos, K.	1943	Lt. Colonel	Infantry	Schoolmaster	Eleochorion
Roufogalis, M.	1943	Lt. Colonel	Artillery	Farmer	Akarta
Stamatelopoulos, D.	1943	Lt. Colonel	Infantry	Farmer	Touzkolegs, Tripolis
Konstantopoulos, G.	1948	Major	Infantry	Unknown	Larissa

Communist position, had an antipolitical bias, and which cherished visions of a "Third Greek Civilization." Surely the Metaxes era left some strong impressions on the minds of the future leaders of the coup. The humble origins of these officers tend to make them suspicious and disdainful of any elite, and help to explain the strain of populism in their post-April 1967 programs, policies, and rhetoric.

If a professional officer corps has its roots in the countryside and tends to come from less socially mobile sectors of the economy, its ideological orientation will be critical of sophisticated upper-class urban values, which it may consider to be corrupt. This attitude may be further reinforced by the professional indoctrination and style of life in the military community, which tends to create hostility toward self-indulgent urban values. If this be the case, then a professional officer corps may well be antagonistic toward a civilian government whose officials arise from different economic and social strata of the society. The military, therefore, is more likely to support a civilian government whose key members have their origins in a social class similar to that from which the military springs.

The next step is an examination of the strata of society from which Greek parliamentarians originate to observe how they compare with those of the officer corps.

Modern Greek politics and life are accentuated by two variables: (1) kinship ties and family connections; and (2) regional loyalty. For most people, the nature of the regime, and even obedience to authority have come to depend on personal ties and clientage networks that connect the individual to particular political incumbents. In this kind of setting actual inheritance of political position is quite common. There are few career opportunities, so political office itself often becomes the basis of the family status and fortune. From a more general perspective, inherited political positions are generally associated with traditional political systems. More precisely, political inheritance tends to be more common in societies with limited role differentiation, and Greece is highly qualified for such a category. Greek politicians are an elite group and one wherein economic, political, and social power "agglomerates." [38] So we may infer that one who comes from a political family is also likely to possess a high socioeconomic status within the framework of Greek society.

A significant number of parliamentarians have emerged from families with histories of political involvement, which might indicate that the network of political influence at the provincial and local levels almost certainly has similar family components.

38. For amplification of this subject see: Sir Lewis Namier, *The Structure of Politics at the Acession of George III* New York: Macmillan Co., 1957); Mattei Dogan, "Political Ascent in a Class Society," in Dwaine Marvick, ed., *Political Decision Makers* (Glencoe, Ill.: The Free Press, 1961).

The data from table 13 indicate that those from political families have retained a strong hold on political offices since World War II. The advantages offered by particular regional backgrounds are related to clientage, and most Greek politicians retain strong ties to their place of birth, even if they later

Table 13: *Family Backgrounds of Greek Cabinet Ministers*
1946–1965

	Father in Politics	Relatives in Politics	Total
Ministers (n 230)	15.6%	21.7%	37.4%
Leaders (n 35)	31.4%	17.1%	48.6%

Sources: *Who's Who in Greece* (Athens: Omiros Press, 1965); Keith R. Legg, *Politics in Modern Greece* (Stanford, Calif.: Stanford University Press, 1969).

live elsewhere. It is apparent from the data in table 14 that Athens is over-represented, atlhough during the 1946–1965 period there appears to be some decline in its ministerial representation. It must be noted, however, that many Athenians had been born elsewhere, and numbers of politicians, like the people that elected them, had migrated to the city. The traditional and historic areas of Greece (Central Greece, Peloponnesos) supply most of the cabinet ministers, whereas Thessaly, Epirus, and Crete, which were recent additions to the territory of Greece (since 1910) were only somewhat slighted during both periods. We can see that Macedonia and Thrace did not contribute any substantial proportion to the cabinet ministries. (It was in these areas, in-cidentally, that the majority of the refugees from the Greco-Turkish wars settled between 1910 and 1936.) Our data for the latter time period continues to reflect this disproportionate representation, offering clear evidence that the refugees and persons born in areas recently added to the Grek state have not as yet filtered to the top of the political ladder. A further explanation as to why these "new areas" have not been able to enter into fuller political participation is the lack of accumulated clientage ties and a corresponding emphasis on personalities and issues. Few deputies from these areas were able to acquire enough political tenure and influence to demand ministerial port-folios. Thus, politicians from the older, more established areas with secure clientage networks, continued to dominate the traditional political groupings.

If we compare the geographic origins of cabinet ministers with those of the officer corps, certain similarities emerge. Central Greece and Peloponnesos contribute a large proportion of the total officer corps, and this compares favorably to these areas' contribution of cabinet ministers. The areas of Crete, Thessaly, and Epirus also contribute a similar measure of officers and cabinet ministers. It should be noted that in the case of Peloponnesos and

Table 14: *Geographic Origins of Cabinet Ministers and Professional Officer Corps, 1910–1965*

	Athens	Central Greece	Pelopon-nesos	Thessaly	Epirus	Crete	Macedonia & Thrace	Ionian Islands	Aegean Islands	Others	Un-known
1910–1936											
Ministers (300)	18.6%	17.0%	22.0%	4.7%	3.1%	3.2%	5.3%	6.0%	0.7%	7.7%	11.7%
Leaders (31)	22.6%	14.8%	14.3%	9.7%	6.3%	6.6%	3.2%	3.2%	—	16.1%	3.2%
Census Data, 1928	13.1%	16.1%	16.0%	8.2%	3.5%	5.4%	31.4%	4.0%	—	3.3%	—
1946–1965											
Ministers (300)	13.5%	16.0%	25.5%	6.5%	5.3%	6.0%	10.8%	4.2%	1.3%	3.9%	7.4%
Leaders (35)	28.6%	18.0%	22.0%	8.6%	4.2%	4.4%	8.5%	5.7%	—	—	—
1916–1965											
Officer Corps											
—Army	10.8%	15.8%	26.1%	7.2%	6.0%	9.4%	15.6%	3.3%	3.8%	2.2%	—
—Air Force	16.0%	24.5%	27.0%	7.0%	2.0%	7.5%	12.2%	4.8%	—	—	—
Census Data, 1961	22.0%	3.1%	13.1%	8.3%	4.2%	5.8%	29.9%	2.5%	5.7%	5.4%	—

Sources: *Statistical Data of the Social Characteristics of Professional Officers* (Athens: Armed Forces Headquarters, Statistical Service, June, 1969); *Politics in Modern Greece*; *Megali Elliniki Engiklopaedia* (1933); *Ellinikon Who's Who* (1965); Survey of 2,000 officers conducted in October 1971.

Central Greece a basic reason for their sizable contributions to both professions is found in their traditional historical roles. These regions, as well as
Thessaly, Epirus, Crete, Macedonia, and Thrace, are relatively poor ones,
economically speaking, and the social mobility, security, and prestige previously touched upon that are offered by careers in the armed forces appeals
greatly to their inhabitants.

One major note of importance is that where similarities do exist in areas
contributing more or less equally to both professions, the politicians and
political families tend to represent the more affluent sectors of these regions;
whereas the recruitment of the officer corps is from lower socioeconomic
strata. Our data also reflects the fact that Athens does not contribute substantially to the ranks of the officer corps, as compared to its contribution of
parliamentarians (about forty-two percent). This is doubtless due to the fact
that it offers wider opportunities for social mobility unavailable in other
regions. It should be pointed out that many of these politicians were born in
areas other than Athens and migrated later to the city, where they no doubt
acquired a certain measure of the more sophisticated, urban-class values,
while retaining their clientage relationships through family ties in the region
of their birth.

The rural social background of the professional officer, coupled with
peasant or middle-class occupational origins, contributes to Janowitz's con-
tention of a "fundamentalist orientation and lack of integration with other
elites, especially political elites." [39] Therefore, since the officer corps has its
roots in the rural areas, its ideological orientation tends to be "fundamentalist" in nature, a moral and religious certainty critical of those from higher
socioeconomic strata, whom they may envy and consider "corrupt and even
decadent." [40] The coup leaders visibly displayed their disdain for urban life,
as reflected in their edicts forbidding long hair for schoolboys, the short-lived
ban on mini-skirts, their insistence that government officials attend church,
the constant plugging of nationalism and Christianity (inclusive of slogans
such as "April 21st—Christ is Risen, Greece is Risen" and "Greece for
Christian Greeks"), and exaltation of the army.

These glimpses into the social origins of the Greek officer corps are somewhat contradictory in that the officer corps is hostile to what it considers self-
indulgent urban values, yet it is oriented to modernization and technological
development. These social backgrounds, together with their military experience, help to make the world of politics easily accessible to the Greek officer,
even though there is a wide cultural gap between the attitudes and ideology of
the officer corps and those of the political leaders. As long as the Greek military profession maintains a narrow social base of recruitment, it will possess
a relatively uniform social and political outlook, which background condi-

39. Janowitz, *The Military in the Political Development of New Nations,* p. 58.
40. Ibid.

tions and fashions conservative political perspectives and sets limits on the links with other elite groups. As a result, these gaps in values and the basic distrust that has long existed between these two elements of Greek society heightened the suspicions of the coup leaders regarding the political chaos of 1965 to 1967. This, in turn, may well have precipitated the April 21 coup d'état.

THE PARADOX OF POWER: POLITICAL COMMITMENT AND POLITICAL DECAY IN FRANCE AND GERMANY IN WORLD WAR II

Edward Feit

Unless we subscribe to the notion of perpetual growth, a notion as difficult to sustain as that of perpetual motion, we need theories of political decay. This was not at all obvious, however, until a hypothesis of decay was raised in a seminal paper by Samuel P. Huntington, who has expanded and deepened it in a recent book.[1] The wealth of insights these works contain, and the influence they have had, make it fitting that some of their fundamental theses be reexamined and, perhaps, expanded and altered. To do so, however, they must first of all be briefly outlined.

Huntington sees political decay as the degeneration of institutions, which he defines as "stable, valued, and recurring patterns of behavior," and the increasing dominance of disruptive social forces. Following Plato, Huntington argues that a corrupt society is one "which lacks law, authority, cohesion, discipline, and consensus, where private interests dominate public ones, where there is an absence of civil obligation and civic duty, where, in short, political institutions are weak and social forces strong."[2]

Political decay is, thus, related to political disruption. Disorder dissolves institutions. As order cannot exist without constraints, it would seem that the maintenance of social constraints is essential to social survival. Social

1. Samuel P. Huntington, "Political Development and Political Decay," *World Politics* 17, no. 3: 416; *Political Order in Changing Societies* (New Haven, Conn.: 1968), pp. 86–87.
2. Huntington, "Political Development," p. 416.

order is, consequently, correlated closely to the concentration of political power in the hands of those who make authoritative decisions.

If an institution is powerful, that is, if it is working well, this implies it is orderly. An institution riven by disorder is not powerful and can act neither to ensure achievement of its ends nor its survival. The assumption of strength in such a combination of power and order would seem to be logically unassailable. But does this logic always hold? And if it does not, then what is more fundamental in determining political decay than disorder? Much of this article is engaged in analyzing these questions. To illustrate the problem, the French army in defeat in 1940 and the German army in 1942, temporarily victorious, are used as illustrations. Using them, we will seek to show that the French army in its moment of defeat had greater power than the German in its moment of victory. If this proposition is borne out, then power and order are not necessarily correlated, and power can reside in an organization in disorder or can be lacking in one apparently working well.

The facts can be briefly stated: in May and June 1940, the French army was tottering to defeat, and its shattered formations, hopelessly intermingled with fleeing civilians, were falling back in disorder. Yet, at that time its commander, General Maxime Weygand, was able, although legally subordinate to it, to impose his will on the civilian government of Premier Paul Reynaud. To the contrary, the German army, victorious to 1942, could not impose even its military decisions on Adolf Hitler who, in 1941, had made himself its commander-in-chief. He followed not the advice, military or political, or his army commanders, but the dictates of his vaunted intuition and the advice of his favorites. The military specialists were neglected and their views largely set at naught.

Is this reversal of power and success unique? If it were, it could well be a special case, anomalous to general political experience. But it is not. Turning back the leaves of the calendar, we find that there was something similar in the events of 1919 at the end of another earlier war. Then it was the marshal of the victorious allied forces, Ferdinand Foch, who was peremptorily silenced by his premier, Georges Clemenceau, when he tried to impose his political views. In Germany, whose armies had been defeated, it was General Wilhelm Groener who imposed his conditions in a deal with the then chancellor, Friedrich Ebert. The arrangement ensured both the survival of the infant German Republic and the autonomy and traditional character of the army.

The similarity of these and other cases to the paradox being examined here can be overstressed. There clearly are differences. For this reason discussion is confined to the later cases from World War II, and even these cannot be treated in depth in a short essay. Focus is on the fascinating paradox of power and potency, which, with its theoretical implications, will be examined in the following pages.

II Some Fundamental Ideas

Before plunging into theory and fact, a few terms need to be briefly defined. Oft-debated terms such as *power* have been used without explanation. This will now be remedied although no "ultimate" definitions will be essayed; broadly acceptable and workable definitions are all that are offered.

Power will be used in a sense similar to that of Max Weber. He speaks of power as the probability of a person or a group of persons carrying out their will against the resistance of others.[3] An alternative definition, which does not clash with Weber's, is suggested by Karl Deutsch, who speaks of an actor having power if he can act out preferred behaviors with least loss of ability to choose other behaviors.[4] Deutsch stresses decision in the absence of new information from the environment. A man or a group with power can act on their definition of a situation, without knowledge of what is happening at the instant the decision is made to act. Power, in these terms also, is closely associated with will, will inspiring the decision. In a similar way Karl von Klausewitz described the fundamental aim of war as the bending of an enemy to one's will.[5] The association of will with power, even in such a brief and incomplete discussion as this, seems well borne out.

Will is itself a complex conception, and can be defined as the *conscious* choice of a goal from among some alternative set of goals. The choice of goal rests on the consolidated preferences and inhibitions of the actor.[6]

If will, as defined, is the mainstay of power, then anything undermining will erodes power. As order involves exercising power, it involves the exercise of will. Erosion of the will to act erodes order. In terms of the Huntington model, therefore, it is not disorder but the *erosion of will* that causes the decay of institutions.

Can this argument be sustained? It can, perhaps, if perceived power—perceived by one in an authoritative decision-taking role—and real power are congruent. This raises the question of the difference between *real* power and *perceived* power where such a difference exists, and it is rare for men to perceive properly their real power. Generally, power is overstated or undervalued. What then is *real* power and what perceived power?

Power, as the term is employed here, is the sum of *all* will and *all* resources available to an actor or a group. It would therefore follow that power is, at any one time, a fixed quantity. The resources, in terms of material goods, available to the individual or the group, can be measured. The will of

3. *From Max Weber: Essays in Sociology,* trans. H. H. Gerth and C. Wright Mills (Fair Lawn, N.J., 1946), p. 180.
4. Karl Deutsch, *Nerves of Government: Models of Political Communication and Control* (New York, 1963), p. 247.
5. Klausewitz, *On War,* ed. Anatol Rapoport (Baltimore, Md., 1969), p. 101.
6. Deutsch, p. 246.

the individual, or the collective wills of the group, cannot. Will, singular or collective, can only be guessed at but cannot be measured. It follows, therefore, that operative power, the power estimated to be present, is more important in the affairs of men.

Whether speaking of *real* or of *perceived* power, will and resources cannot be divorced from it. If only will mattered, without connection with resources, a Hitler trapped in his beleaguered bunker and ordering nonexistent armies to his rescue, could prevent the destruction of his regime. If only resources mattered, a wealthy Stoic would have immense power despite his voluntary deflation of will. Power thus has two facets: will—the ability to choose a goal without reference to immediate information, and resources—the material means by which to effect the decision to take action.

Organization itself, or even institutions, now need description. Organization clearly is a resource by which the will of the group is manifested. It is, however, not an essential resource. An unorganized group, such as a mob, may face considerable physical opposition without being checked in its path, and may be able to make its unarticulated will unmistakably felt. On the whole, however, some form of organization is generally involved in questions of power.

Power, explained here as having a strongly subjective element, is a matter of intangibles as well as tangibles. It is this that makes the conception elusive without reducing its reality. The question, for the political actor or actors, is where the limits of power are. Here we can speak of *limit* images as defined by Deutsch and assess their importance in the exercise of political will. Limit images are the perceived social and physical limits constraining an actor, associated with the probabilities of encountering them under given conditions.[7] These limits are estimates only, for they have no direct quantitative referent. The "manipulation of risk" in present-day international politics, based on an image of resolution, has its counterpart in all politics. Estimates have to be made of the likelihood of action, the exercise of power, generating overwhelming internal or external resistance.

As the exercise of power involves pushing will to some limit in opposition to other wills, it always involves costs. In systems with highly legitimated constraints, with well-developed ritual of politics, costs may be low. In systems where political ritual has lost its self-affirming purpose, where constraints are new or not yet fully legitimated, and where resources are limited, costs can be high.[8] But they must be either high or low, as they are inescapable. The costs of power can be measured in the level of resources that have to be committed to maintain an act of will. If resources are invested unwisely, resources, and consequently power, can be lost. Power can be lost, in other

7. Ibid., p. 212.
8. Gerhard Lenski, *Power and Privilege: A Theory of Social Stratification* (New York, 1966), pp. 54–58.

words, if the resources that would enable an individual or group to prevail over others is misapplied or dissipated. While misuse or dissipation of resources is one way of losing power, another is to commit no resources, or to commit insufficient resources, to make the actor's will prevail when a value-position reinforcing the image of power is threatened. Power can be lost because those wielding power fail to make the necessary commitment either through neglect or through fear. They see signals heralding the approach of a social or physical limit before the real limit is reached.[9]

This argument brings to the fore the vital question of commitment. If power is the sum of all will and all resources that can be made effective at a given time, then *commitment* is the volume of resources that an individual or a group is prepared to invest in the service of a selected value-position. When an individual actor or group invests *all* resources, we can speak of a total commitment. The general who risks his entire force in a single battle, is totally committed to winning that battle. It is that one and no other. It follows, therefore, that the level of commitment less than total, is indicated by the quantity of resources to be committed willingly. Were resources in politics purely material, then we would have a ready index. Unhappily, this is not so. For, in addition to physical resources, we must add symbolic resources, such as statuses, credibility, and so on, which are subjective. These subjective resources can be committed just as can physical ones, as for instance, when a politician "lays his reputation on the line." Levels of commitment in any situation differ, because estimates of what is to be gained against what must be invested differ in different actors. With power as with commitment, we are dealing with an intangible reality.

Commitment differs from power, and is related to the level of power to be applied in any instance. Commitment, unless it is total, will depend greatly on information received by the actor at the time resources are to be invested. The difference may be made clearer. The power of a sports car and driver as a "man-machine-system" is the maximum speed the car can attain. The perceived power is the speed at which the driver believes the car will travel. His commitment is the actual speed, given road conditions, at which the driver will travel.

The argument has now led to the major point to be made in this essay: political decay is a consequence of a loss by authoritative actors of a willingness to commit resources in defense of value-positions vital to institutional maintenance. The decision-makers fear to exercise power even if the institution will fail without this exercise. Institutions decay because of a *loss of commitment*. Loss of commitment can be described as containing the following elements: willingness to invest resources is inhibited by insecure knowledge; by fear of the consequences of action; by the fear of losing resources

9. Deutsch, p. 246.

in hand; and by the hope that things will, if left alone, get better of their own accord without risks having to be taken.[10]

It would clearly be foolish to ascribe the decay of institutions to one cause alone. Yet, whatever the other causes, loss of commitment would seem to be a highly significant cause of political decay as the cases in point will, it is hoped, illustrate. The cases of the French and German armies set out below bring forth no new material or new presentation of the facts. The very fact that this is so strengthens the argument developed in the preceding pages. Why, it can now better be asked, did Maxime Weygand, head of a bankrupt military gerontocracy wield so much power over the elected representatives of France, men whose instrument he was both by law and custom? And why did German generals, heirs of a proud tradition of service and autonomy, fail to assert themselves against the former corporal Adolf Hitler?

III The Cases: France

By the end of May 1940, the French army was in danger of complete dissolution, and to make matters worse there was little to be hoped for from her ally, Great Britain, or from the neutral United States. What was to be done? Should France capitulate? Should she surrender her army and continue the war with her navy and the troops stationed in her vast colonial empire? Or should the French army refuse to surrender and be evacuated to North Africa to continue the fight? These were both political and military questions, but the primacy was political, and the decisions taken would affect the future of the French state. Were either of the last two alternatives—the surrender of the army in the metropole without an armistice, or the evacuation of the army—to be adopted, it was certain that the whole of the metropole would be occupied by German forces with all the attendant suffering this would inflict on the citizenry. Was this acceptable to the French government? These were the painful alternatives, and it was in the choice among these alternatives that Weygand's became the deciding voice. Understanding why, at this fateful hour for France, he was able to do so calls for an understanding of the causes of the fall of France.

The fall of France is a much debated subject, and there is room for little more than hints of what happened and what the argument is about. The more superficial view is that France, faced with the "overwhelming superiority" of the German army, was reluctant to face another bloodletting for the second time in a century. There is, of course, some truth in these arguments, but they fail to account for the whole picture, for the defeat of France, as recent historians have shown, was far from a foregone conclusion;

10. This passage is adapted from Edward N. Peterson, *The Limits of Hitler's Power* (Princeton, N.J., 1969), p. xvii.

their belief is that faulty military doctrine together with widespread defeatism, rather than any clearly defined shortfall in men or weapons, led to the downfall.[11] As in much political speculation, much of this argument must be "contrafactual"; it must deal with events that did not happen but might have happened had different decisions been made. Such "contrafactuals" are difficult to establish but are essential in political analysis. The secondary sources used will be employed to show the feasibility of the contrafactuals, though acceptance or rejection of them rests with the reader.[12]

Was France in fact far behind Germany in arms and men? A case can be made that France was as well as or even better equipped with modern arms as was Germany. France had more tanks and, if part of the tank force was obsolete, so was that of the Germans. The French *Somua* and *Class B* tanks could match anything that Germany could put in the field except perhaps the Panzer V, and French artillery although lacking heavy guns had the best of the lighter pieces, the famed "75." Faulty military doctrine and the way these resources were used, made defeat sure, for, instead of massing tanks into regiments, divisions, and brigades, she distributed them in "penny packets" among the infantry.

Indeed, France entered the war determined to hoard her resources and to suffer as little loss as possible. The holocaust of 1914–1919 was seared into Frenchmen's minds, soldiers and civilians alike, and none wanted a repeat performance. It was widely believed that France could not again sacrifice one-quarter of her youth and survive as a nation. The costly mystique of *attaque* that had dominated military thinking in the years before 1914 and during the war had been abandoned after it. Static defense behind a fortress line was the answer. If France were "walled off" from Germany, the argument ran, her army could sit out the war waiting for the blockade to exhaust the enemy. If attack became necessary, and this was not ruled out, the French army could invade Germany through Belgium, an ally, and thus run a kind of reverse Schlieffen strategy.

Once adopted, this doctrine acquired the kind of sanctity that such doctrine usually assumes, even after it comes to fly in the face of fact. Belgium opted

11. Recent writings have laid to rest the hoary myths of allied unpreparedness and German preparedness. The question is not whether the Allies were prepared or not, for they were quite well prepared, but for what they were prepared. See for instance, Colonel Adolphe Goutard, *The Battle of France 1940*, trans. A. R. P. Burgess (New York, 1959), pp. 13–16, 23–44; and Alistair Horne, *To Lose a Battle: France 1940* (Boston, 1969), pp. 182–84; Paul Marie de la Gorce, *The French Army: A Military-Political History* (New York, 1963), pp. 297–98; and on the German side with a somewhat different interpretation, General Siegfried Westphal, *The German Army in the West* (London, 1951), pp. 83–84.
12. Abraham Kaplan, *The Conduct of Inquiry: Methodology for Behavioral Science* (San Francisco, 1964), p. 91.

for neutrality at the outbreak of war, without this changing the French conceptions, although in pursuit of neutrality Belgium refused to coordinate her
plans with those of France. Yet, even then, allowing for all the faults of
French planning, the outcome was not decided. Had the French attacked
while the German army was fully engaged in Poland, they might well have
prevailed.[13] The German army was far from perfect then, and harried both
by its own troubles and the enemy. As things turned out, Poland was the
proving ground on which Germany perfected her techniques of *blitzkrieg*.
These bore fruit when France was attacked in turn. The German army
launched an offensive early in May 1940, which soon turned into a rout. The
offensive and its outcome is too well-known to require restatement. During
the debacle the French commander, General Maurice Gamelin, was replaced
by General Maxime Weygand, who, after spending a few precious days
reviewing the situation, set up what he termed his "final" defense line.

Weygand, in setting up his line and stating the condition, made a political
as well as military decision. In permitting Weygand to do this, the French
government reduced its alternatives to the first three described above: capitulation of the government; capitulation of the army and the government
going into exile to continue the war; or an evacuation of the army with both
government and army continuing the war from abroad.[14]

One estimate was vital in determining the choice: would Britain continue
the war? A move to the colonies would be advantageous only if Britain was
able to sustain her war effort, and if ultimately America threw her might into
the scales. If, soon after France was abandoned, Britain was defeated in turn,
there would have been no advantage and only disadvantage in continuing the
war from exile. As things stood, France still had her fleet intact and her empire, so Germany still stood to gain advantages from a French surrender,
advantages which would vanish with the defeat of Britain. France would,
then, lie naked before the might and the vindictiveness of Hitler. Estimates of
power among individuals, perceptions of limits—personal and social, these
would govern alike the acts of Reynaud, of his cabinet, and of the generals.
Here the differences were profound: Reynaud believed that Britain would
resist successfully and that France should continue the war; Weygand and
his confreres did not believe that Britain could long survive the defeat of
France.[15] The military will prevailed.

Cutting rapidly through a maze of issues such as this is to explain little.

13. This argument is worked out in John Kimche, *The Unfought Battle* (New
York, 1968). The argument is supported from German sources by Horne, p.
100, and Westphal, p. 71. According to the latter, who commanded a unit in
Germany at the time, French inaction seemed inexplicable to the Germans.
14. On the logic of the armistice, see Paul Kecskemeti, *Strategic Surrender: The
Politics of Victory and Defeat* (Stanford, Calif., 1958), pp. 31–34, 40–41.
15. de la Gorce, p. 299; Kecskemeti, pp. 36–37.

The question remains: Why did Weygand get his way? How did he do it? What, in short, was he willing to commit and what did his civilian opponents fail to commit? Here, after all, was the commander of the defeated army setting his limit images far beyond those of an elected government, and able to enforce his will. How? Clearly no more than partial answers can be offered, but even these are illuminating. The picture can not be fully explained because of the conflicting claims of the principal actors and others, but enough observations can be made to allow for some meaningful generalizations.

Weygand's power rested on his will and on an array of resources. His will was given direction by his conceptions of the French army, the French nation, and the nature of the representative regime in France. His resources included his expertise, the coherence of the body supporting him—the other French commanders, who saw the world in much the same way as Weygand and did not greatly differ from him in their estimates of what ought to be done. Their values approximated each other's closely. The same could not be said of the political leaders whose perceptions were more incoherent, who were not united on the most essential issues, and who bought at least part of the army's perception of itself.

Weygand, in common with his peers, saw the army as "the true France"— as an integral body, a nation whose unity was disrupted and obscured through the machinations of politicians and parties. It was these that had weakened France by their partisanship and had denied France and her army the weapons they so badly needed to defend the country. They had undermined and scorned patriotism and denied to the patriotic the means by which the *patrie* could be saved. The politicians alone were to blame for the defeat and should be made to shoulder their guilt. It was the politicians, inside and outside the government, Weygand had said, who had "allowed twenty-two years of political illusion, unpreparedness, antimilitarism, and antipatriotism," to persist. Nor should the cogently expressed view of Marshal Petain in a famous broadcast of June 20, 1940, be forgotten. In it Petain said that "too few children, too few arms, too few allies" were the causes of the downfall of France. From this set of causes it followed that no part of the defeat of France could be laid at the door of the army, and any effort to do so was part of a diabolical plot on the part of the politicians.[16]

The French government, then, was directly responsible for the French defeat, but what about the French people? The French army identified not with living Frenchmen and Frenchwomen, but with an abstract French "nation" existing apart from its citizens. Indeed, judging from their statements,

16. Philip C. F. Bankwitz, *Maxime Weygand and Civil-Military Relations in Modern France* (Cambridge, Mass., 1967), pp. 210–11, 315–18; William Shirer, *The Collapse of the Republic: An Inquiry into the Fall of France in 1940* (New York, 1969), pp. 807–10, 817–18.

they had no high regard for their fellow citizens. The officers, identifying with a mystical and symbolic France, believed themselves capable of its regeneration, once it rejected the politicians and their divisive ways.[17]

If the army saw itself as the embodiment of the nation, then what did the politicians represent? To the generals, the politicians represented only their few followers united under a party label—a point Weygand, for one, never hesitated to make. In one impassioned display of rancor in the presence of witnesses, he challenged Reynaud's right to speak for France. At another time, when Reynaud cited the queen of the Netherlands as an example of a head of a government going into exile, Weygand drew invidious comparisons between the queen and the French premier as national symbols.[18]

The belief that civilian politicians enjoyed a broader mandate than the army was weakened among soldiers well before the German armies struck. In the interwar years the officers had perfected techniques enabling them to intimidate civilian politicians while paying lip-service to the politicians' superordinate role. This had a double-edged effect: on the one hand, the soldiers, despite their independent stance, were still enmeshed in the myth of the army as "la Grande Muette" and as such bound to obey civilian authority. They therefore expected that, no matter how reckless their behavior to politicians, the government would call them to order in the last resort, saving them from the consequences that would follow if their (the soldiers') demands were met. The government would rescue the soldiers at the last minute, they hoped, and thus they could wear the mantle of heroes without risk. Weygand seems to have relied on something like this when obdurate, and was disappointed when the government was unable in the last minute to exert what he felt ought to be its will. Like the reckless youth of fiction, he hoped to be allowed to gamble, seemingly committing his all on the turn of a card, while all the time secure in the knowledge that his father would, in the end, pay his debts. He seemed unaware, or did not wish to be aware, of the extent to which army pressures had hollowed the authority of the government. The civilians could not exercise their powers even if they wished to do so, for failure to act over so many years had eroded their moral capital. It was not incapacity that inhibited the civilians in commanding soldiers, but the fact that their powers had virtually vanished vis-à-vis the army.[19]

Paul Reynaud is agreed to have been a man of personal courage, patriotism, and of no mean political skill. Yet, for all that, he was not a wartime leader of the order of Danton, Gambetta, or Clemenceau. There is much in Weygand's harsh judgment of both Reynaud and of himself as "unworthy of

17. Bankwitz, p. 292; Shirer, p. 808, 814–15; Kecksemeti, p. 41.
18. de la Gorce, pp. 293–7; Bankwitz, p. 297; Horne, pp. 116–22; and for a German view, General Ulrich Liss, *Westfront 1939–1940* (Neckargemuend, 1959), pp. 46–47.
19. Bankwitz, p. 320.

the great ancestors." [20] But in saying this one must consider the contrafactuals. What commitments *might* Reynaud have made? What, given will, were his resources?

The powers of government had been whittled down by the time Reynaud took office, and he took over the government when the country was already on the road to defeat. Were any options open to him? There were some: He could have used his constitutional prerogative to withhold approval of Weygand's "final line." He could have then demanded that his commander-in-chief prepare a plan consistent with the government's policies. Had Weygand refused, the plans could have been formulated by officers sympathetic to Reynaud's stand, by men such as Colonel Charles de Gaulle or Colonel Villelume. Reynaud might have called a meeting of the *Conseil de Guerre,* which would probably have supported him. He did none of these things. Why? What prevented him from playing the cards he held?

Later, after the events, Reynaud was to offer a number of explanations and excuses—none of which quite hang together. Yet the issue was whether the civilians or the military were to determine the conduct—political and military—of the war. On the crucial day, May 25, Reynaud was, it seems, trapped in a difficult and dangerous position, and this blinded him to the issues at stake. Four days later, when it was too late, he realized what was at issue and began challenging Weygand's decisions, but the die was cast.[21]

Many things prevented Reynaud from committing himself in his struggle with Weygand. Among them was the blurred line between "grand strategy," or the adjustment of the conduct of war to political and military realities, and what can be termed "pure strategy," the conduct of battlefield operations. What is the civilian and what the military sphere in a war involving an entire nation? How are these spheres to be combined if they can be considered separately at all? The blurring of functional lines led the civilians to accept the army at its own valuation, and to see themselves much as the army saw them—though they knew the military to be biased and pejorative.

The effectiveness of any commitment by Reynaud, to what he saw himself as able to commit, depended on his view of his mandate. Reynaud could not base his authority on the ballot box as could Churchill. His government was the product of the usual French political carpentry.[22] It came into being not at the behest of the nation but because of a series of political deals. It comprised a kaleidoscopic spectrum of opinion from "hard" to "softliners" on the war. It had some who favored fighting to the finish and others ready for peace at almost any price. Between the extremes were others who trimmed their sails to the political winds. Yet, there is evidence, albeit arguable, that,

20. Ibid., pp. 303–5.
21. Ibid., 298–9.
22. de la Gorce, pp. 290–1; Bankwitz, p. 311.

in the end, a sizable majority would have followed Reynaud's lead.[23] Had
he opted for the North African solution, he would very likely have won his
way. This leaves the unanswered question: Why did he not make the commitment and risk resources he obviously commanded?

The effectiveness of an exercise of power by Reynaud depended on his
perception of his resources. Being unable to say that he could command a
majority of Frenchmen, Reynaud had to base his claims to speak for France
on something more subjective and intangible. He seems to have doubted
that, were he to go into exile, the French people would still recognize him
as their spokesman. He doubted if he could convince the people of the wisdom of a move to North Africa once Weygand's final line was breached, as
it was bound to be. A move to North Africa was, after all, opposed not only
by Weygand but also by the aged Marshal Petain, a living legend wrapped
in the mystique of Verdun.

There were, in addition, stubborn objective matters mixed in with the
subjective. The government seems to have believed that it could not assert
its prerogatives against the army in time of war, and that asserting them
would only increase the division and panic among the French. Weygand,
called on to resign, might refuse. The government would then have to remove him, setting off a scandal, and further dividing army, government, and
nation. The government would be faced, at a time of crisis, of replacing
Weygand from among the other aging generals. The successor would have
to be acceptable to his brother officers, and there was doubt as to whether
such a man was to be found.

There was, in addition, the French state of mind to be taken into account.
The news of a decision to evacuate the government to North Africa, and the
loss of the *metropole* this represented, would add millions more to the refugees already streaming southward. Yet French opinion might even then have
been reoriented had politicians and soldiers spoken with one voice, but this
was impossible. Weygand was set in his opposition to any move to North
Africa; the move would have been possible only were Weygand removed
from command. The attempt might have angered senior officers and have led
to the resignation of Marshal Petain, whom many Frenchmen saw as their
savior, and this act would have eroded what little spirit the country still possessed.[24]

Running the gamut from the decorous to the positively rancorous, the
conflict between Reynaud and Weygand ran its course, punctuated by feeble
efforts on Reynaud's part to recapture lost initiative with moves such as
"suggesting"—but not ordering—retreat to the "Breton redoubt" as a step
to North Africa. The question was academic by June 13, for the possibility

23. See, for instance, Shirer, pp. 803–21; Bankwitz, pp. 300–1.
24. Bankwitz, pp. 302–3; de la Gorce, pp. 303–6.

for asserting civilian control was long past. It was Weygand who "demanded, enforced, imposed" the armistice resolution on a by then frightened and irresolute Reynaud. The premier could only make the empty gesture of resignation in favor of Marshal Petain, although it could again be argued that, had he forced the issue, the majority of the cabinet would have followed him.

By committing his resources, Weygand had won out. He risked removal from command and possible discredit. But his refusal to leave France, his determination to surrender his army if allowed to remain, and his certain though unspoken refusal to resign had enabled him to dominate the scene. Yet, had Reynaud made *his own* demands, it is likely that he would have won out.[25] General Huntziger had been sounded out and seemed willing to assume overall command. Most of the officers, in terms of a somewhat shopworn but still useful tradition, would probably have followed a determined civilian leader. Brute courage would have been required of any politician in this role, but civilian control might have been successfully reasserted—at least in the early days of May and even in June. But Weygand was not effectively challenged, and he obtained the armistice he had so much desired. This led, in its turn, to the four-year occupation first of part of France and later of the whole country. It also led to the rise of a young officer, then unknown, Charles de Gaulle, who was, at considerable risk, ready to commit all to a French victory. It was this readiness for commitment that enabled him to assert French rights when French fortunes were at their lowest ebb and to succeed against the many opponents who wished to replace him. This, however, is another history which cannot be detailed here.

The shared perceptions of French politicians inhibited them in their dealings with soldiers, in turn preventing them from committing resources at crucial moments in the face of military pressure. They were defeated for, in a broad sense, political relationships depend on conflict, and the one who wins is the one willing to commit more to the struggle.

In assessing political conflict, and the role of politicians and generals, one can see that the threat of the use of resources is as important as their actual use. Decisions are taken on estimates of an opponent's commitment, which is compared with the commitment the actor himself will make. The result can well be based on the estimate of the "worth" of the goal each side has, and this worth may be objective or subjective, or it may be a symbolic goal or a material one. Yet overawareness of the implications and costs in pursuit of a goal may be as harmful in conflict as ignorance. Overawareness of risk may reduce a willingness to commit resources to conflict. To say this is not making a virtue of stupidity. It is rather to say that men who perceive reality in narrower terms, whose values are "peaked" in terms of priorities, can make greater commitments than men of broader outlook.

The peaking of values can be an advantage in times of crisis as it presents

25. Shirer, pp. 833–43; de la Gorce, pp. 306–8.

alternatives in clear order of choice. But, in noncritical times, where issues can be contemplated and where decisions are based greatly on persuasion and consensus, it is men of broader view who have the advantage. Able to see many alternatives in each issue, men of broader vision begin to see too many such alternatives at a time of crisis, perceive the enemy as too subtle, strong, and tenacious and as a result tend to act with action where impetuousness would be better, to hesitate where to hesitate is to lose, and to delay where to delay is fatal. Men of more limited perceptions commit resources more recklessly and in so doing may sometimes prevail in times of crisis.

The dichotomy outlined is to demonstrate archetypes rather than to describe what is found in nature. Nonetheless, there is some analogy in the roles of Weygand and Reynaud. Indeed such archetypes, in modified form, are found even in the annals of the French army. Both Maurice Gamelin and his successor, Weygand, were intelligent men. Gamelin was an intellectual in the true sense, steeped not only in the practice of arms but in French culture. Weygand was narrower, a product of Jesuit training, and with his sights set firmly on military science. There can be little doubt who, at the time of the fall of France, was more willing and able to make commitments and thus to gain his aims.

The argument can be set out in terms of the contraction of psychological time and its effects on the men of broadly spread or peaked values. In times of crisis psychological time contracts. The man who sees too many alternatives consumes too much chronological time in arriving at a decision. The narrower man, perceiving fewer alternatives, can choose among them more quickly. Changes in psychological time, as distinguished from chronological time, thus change the ascendancy of the archetypical leader. The men of broader perception tend to become incapacitated when there is insufficient time for contemplation, the men of peaked values when contemplation must precede decision.[26]

IV The Cases: Germany

The French case allows the conflict to be personified in its protagonists and to be presented as a conflict of this kind. It therefore runs the risk of being, to some extent, a variant of the "great man" theory. As men with leadership capacity do unquestionably influence the course of history, this

26. An intuitive definition of the contraction of psychological time is to be found in Arnold Kaufman, *The Science of Decision Making* (New York: 1968). Kaufman describes the phenomenon in the following way: "Let us call 'fact' or 'event' a modification, whether under the influence of environment or under the influence of decision, of the system studied; in that case the same period of time will be richer in events for the later periods. If instead of dividing time into equal intervals, we divide it into intervals each containing the same number of facts, the duration of each of these intervals will be smaller and smaller" (pp. 19–20).

is not necessarily an evil in itself. However, the German case allows little for personification, other than the demonic figure of Hitler himself. The officers can, despite their differences, be anonymously aggregated as "the generals."

There are several reasons for this. The first is that the German officers, as a class, cultivated the rule of anonymity. The second is that no truly outstanding personality emerged from among the officers with undisputed marks of the leader. After the resignation of Colonel-General Hans von Seeckt in 1926 no one man can truly be said to have commanded universal influence and authority. General Kurt von Schleicher, of course, was a clever political manipulator, but hardly characteristic. Wilhelm Groener, another officer of stature, had resigned from the army in 1919. The men who later conspired against Hitler, in 1938 and 1944, including even the noble-spirited Ludwig Beck, emerged later and seem to have lacked substance and solidity despite biographers and hagiographers. Certainly Werner von Blomberg, aptly nicknamed "the rubber lion" by his colleagues, hardly matched the fiercely conservative Weygand, fateful though Blomberg's role was to be in bringing the army into subservience to Hitler in the thirties. Nor can much be said for Blomberg's colleague of that time, General Werner von Fritsch, who was in his turn nicknamed "the sphinx without a riddle." [27]

Perhaps most of the German higher officers seem to have been so insubstantial because they became jackbooted clerks—bureaucrats with arms who fully subscribed to the bureaucratic ethos.[28] In case this judgment seems unduly harsh, one need only to look at their conduct. As Gordon Craig points out, the generals continued to fight for Hitler even though they had lost confidence in his conduct of the war, despite the fact that his intervention in technical military matters violated their professional standards, and despite the fact that their so doing cost lives needlessly. "They fought on while they considered whether they should join the conspiracy; they fought on while a new wave of proscriptions passed over the officer corps; and they were still fighting hopelessly amid the ruins of their country when Hitler put a bullet through the roof of his own mouth in the bunker in Berlin." [29] If this is not being clerks, it is difficult to define what is. The generals largely did as they were told, though they stood around in corners complaining that

27. For assessments of von Blomberg and von Fritsch, see, for instance, Friedrich Hossbach, *Zwischen Wehrmacht und Hitler* (Hanover, Germany, 1949), p. 76; Gordon Craig, *The Politics of the Prussian Army 1640–1945* (Oxford, 1955), pp. 489–96; John W. Wheeler-Bennett, *The Nemesis of Power: The German Army in Politics 1918–1945* (London, 1953), pp. 295–7, 301–4; Robert J. O'Neill, *The German Army and the Nazi Party 1933–1939* (New York, 1966), pp. 16–17, 24–30.
28. The conception of soldiers as armed bureaucrats is developed in E. Feit, "The Rule of the 'Iron Surgeons': Military Government in Spain and Ghana," *Comparative Politics* 1 (July 1969): 485–97.
29. Craig, p. 502.

things were not as they should be.[30] The few outstanding characters among the anti-Hitler resistance, and, taking the totality of officers and civilians in Germany there were embarrassingly few who joined, were far outnumbered by those who planned to join as soon as success was assured.[31] Petain and Weygand may have been wrongheaded and perverse, especially in the light of hindsight, but they breathe more of life and individuality than do their German counterparts.

It is almost a truism to point to the very different traditions and styles of the German and French governments. Yet, in the thirties at any rate, the German army, like the French, could have brought crushing pressures to bear on Hitler and at different times even had unmade him. The fact that Hitler was compelled to make deals with the generals at all, such as those to be mentioned below, taken together with his careful treatment of the army before 1938, points to this state of affairs. The significant question then is: What eroded this power? Why was the army, which had had such immense resources of political power in the days of the Weimar Republic, rendered impotent against a Hitler at a time when its expertise was so manifestly evident?

Comparison of the German and French cases has many fascinating aspects, not least of which is the way they each seem an obverse of the other. The army, by bringing a variety of pressures to bear in France, hollowed out the civilian regime while continuing to pay it rhetorical deference. In Germany, on the other hand, Hitler imposed an increasingly rigid hold on the army while paying lip service to its traditions and importance to the state. His hold consolidated in the end, Hitler subjected the officers to a series of humiliations which no previous ruler would have dared to inflict on those who wore the uniform. Officers were made to violate cherished traditions, to violate their code of honor, and—swallowing their pride—to suspend their personal, professional, and political judgment.[32] Its will to resist undermined by the dismissal of the pliant von Blomberg and the silent Fritsch, Hitler "threw aside the facade and, in a series of brutal maneuvers, arrogated to himself personal command of the armed forces." [33] This action was not challenged by the army leadership and marked the beginning of a road most officers followed to the bitter end. A road that bound them indissolubly to Hitler's crimes. Hitler destroyed the commitment to professional values among his higher officers by first flattering the army, and then by inducing them to accept a series of compromises which appeared to favor the army while actually undermining its power.[34]

30. General Heinz Guderian, *Panzer Leader* (New York, 1952), pp. 349–50.
31. Wheeler-Bennett, p. 695.
32. Craig, p. 469.
33. Ibid., 469–70.
34. Ibid.

While it is again impossible to enter fully into the history of the time, and this analysis does not require such an exercise, certain issues stand out in the erosion of commitment, and the subsequent decay of the German officer corps. One such crucial issue was the question of succession to the ailing and aged president of the Reich, Field Marshal Paul von Hindenburg. Hitler was eager to succeed Hindenburg and knew, as did the army commanders, that the permission or, at least, the neutrality of the army, was essential to this goal. The army, in its turn, needed Hitler. The brown shirted legions of the S.A. (Sturmabteilungen), with their commander, ex-Captain Roehm, as their loudest voice, were demanding a share in the spoils of power. More specifically, they demanded incorporation in the army, lock, stock, and barrel, with their S.A. ranks translated into their military equivalents. The army officers viewed this with horror. The brownshirts were a rowdy, brawling, mainly lower class body, who, in the eyes of the soldiers could only be incorporated into the army on an individual basis. Blomberg therefore told Hitler, on behalf of the officer corps, that his candidacy for the presidency would be upheld (and thus virtually assured) if the S.A. were crushed. When Hitler hesitated, he was warned that, in terms of the still extant constitution, Hindenburg could (and would) declare martial law. These arguments seem to have persuaded Hitler. The power of the S.A. was broken in the bloodbath of June 1934 and the S.A. rendered innocuous. Hitler became president and remained chancellor, both offices later being combined into that of *Fuehrer*.[35]

The army, events were to prove, had made a bad bargain. Once the offices of president and chancellor were combined, no appeal could be made to another source of power. Even the destruction of the S.A. leadership and its cowing by the S.S. (Sturmstaffeln) meant only, in the end, that the S.S. became a rival of the army for the cherished role of "sole arms bearer" in the state. The army was, after all, not to monopolize the profession of arms. They could not meet the challenge of the S.S. which in time became a parallel army with all arms of its own. They lost their prized autonomy and the weight that their voice had had in the councils of the state.[36] The army slowly lost control over its fate, with neither the will nor the resources to change its course in a more desirable direction.

35. There was an earlier agreement between Hitler and von Blomberg as well, which facilitated Hitler's appointment to the chancellorship. See Francis L. Carsten, *Reichswehr and Politics 1918–1933* (Oxford, 1966), pp. 393–4. For the later agreement and the destruction of the S.A. leadership, see Westphal, p. 9; Demeter, pp. 201–5; O'Neill, chap. 3; Klaus-Juergen Mueller, *Das Heer und Hitler: Armee und Nationalsozialistische Regime* (Stuttgart, 1969), chap. 3.

36. Mueller, p. 136. Karl Demeter, *The German Officer Corps in Society and State, 1650–1945* (New York, 1965), p. 207; Westphal, pp. 46–48; Craig, p. 470; Wheeler-Bennett, pp. 290–1, 333–8.

Among the lost resources was one common to the German and the French armies: the notion of the identity of army and nation. The German army, much as did the French, saw itself as the true embodiment of the mythical German people. The fall of the Kaiser in 1919 had removed the army's focus of loyalty, and it was replaced by the idea of the nation. Governments in Germany, as in France, were seen by officers as representing not "the people" but the politicians who had put them together and their followers. Loyalty in Germany was owed not to the governments with their shifting coalitions but to the German "Volk" and to the state.

In this set of beliefs the German officers at first enjoyed an advantage which was denied their French counterparts. German professors, in the nineteenth century, had been busily fabricating an ideology of "Volkstum" replete with racial communities transcending any mere agglomeration of mortals. This body of doctrine had been widely adopted in Germany, and all the army needed in the years before Hitler was to pick up this already respectable body of doctrine. The fall of the emperor and the German princes had cut off the apex of the pyramid of power. The army turned to its base. The army looked to replace the special relationship it enjoyed with the rulers in two ways: first of all, by comforting itself with a series of pseudo-rulers such as von Seeckt, Hindenburg, and finally Hitler; and second, because this was unconvincing to the majority, by tying themselves to the "people." The army could represent that mystic entity, the "Volk."

This second prop was, however, pulled from under the officers by Hitler. How could the army claim to represent this "Volk" when the Nazi Party claimed to do so, and its claim seemed far more effective? The Nazis had come to power claiming to personify the German nation, the spiritual unity of Germany. How could two bodies personify what both agreed to be a spiritual unity? The Nazi party was able to make its claim hold all the more strongly when the army was diluted by the inflood of conscripts after 1935, many of whom had been strongly influenced by the party. The notion that the army embodied the nation had to be reluctantly abandoned. This weakened the resolve of the officers who felt a lack of sympathy for Nazism. Having accepted the mystique of the Volk and having half-accepted the identity of Volk and Nazi party, the officers were forced to accept that the Nazis and not they represented the nation and incorporated its spirit.[37] They were forced back to the apex of the pyramid and the oath to Hitler again gave the corps its focus.

37. On the nature and importance of the oath see Mueller, pp. 134–141, Gert Buchheit, *Soldatentum und Rebellion: Die Tragoedie der deutschen Wehrmacht,* (Rastatt, Germany, 1961), pp. 38–70, Karl Demeter, *The German Officer Corps in Society and State 1650–1945* (New York, 1965), pp. 206–213, Craig, pp. 479–481, O'Neill, pp. 54–58, Wheeler-Bennett, pp. 394–395. Demeter, pp. 59–60, 191; Carsten, 400–1; Westphal, p. 8.

Hitler had a number of cards to play with the officers, his highest being the plan to reintroduce conscription denied by the Versailles Treaty. This expansion would not only enhance the security of Germany and give her a greater voice in the councils of nations, but would also mean more rapid advancement in the army and the adoption of modern military technology. Conscription thus served the principled and expedient interests of officers. Expansion did involve costs for the army, for the rigid principles of selection which had operated in 1935, the year before conscription, would have to go by the board. The ranks of the armed forces could not but be filled with Nazi supporters or sympathizers, and this was true not only for the young and enthusiastic junior officers but also of the "retreads"—officers who had found no place in the old Reichswehr and were now brought back to fill out ranks in the new Wehrmacht. The new men politicized the army and canalized differences within the officer corps.[38]

The coherence characteristic of the corps before 1933 and which had been among its most important resources began to disappear with the new influx. For different reasons each segment into which the corps was increasingly split served for different reasons and reacted differently to Nazi threats to value positions. Some continued to serve for reasons of patriotism and out of a desire to avenge the defeat of 1919. Others found Nazism to be congruent with their own hopes and aspirations. Others, again, such as General von Reichenau, disguised what was suspected to be naked careerism behind euphoria for Nazism. In many others all these combined in different measure. Be that as it may, the officer corps which, at one time, had been able to confront the politicians in a phalanx was now divided within itself. The patriots, the technocrats, the careerists, and the opportunists spoke with different voices and pulled in different directions. This allowed Hitler the role of arbitrator among the different factions, and gave him power over them all.[39]

Resources had been lost—the identity with the "Volk" and professional unity of the corps, but another remained: expertise. In a conflict over the murky boundaries of grand strategy they might yet have tipped the scales in their favor by throwing their skill in the scales as the French army had done. It did not work in Germany. The reason was, perhaps, the very rapidity with which Hitler won victories on the international scene, whose possibility the soldiers had doubted. The overwhelming character of these successes stunned Hitler's military critics, when, against their advice and in the face of their fears, Hitler remilitarized the Rhineland in 1936, occupied

38. O'Neill, pp. 56–57. Craig, pp. 481–4; Demeter, pp. 57–58, 199; Westphal, pp. 38–39.
39. Carsten, p. 399; Wheeler-Bennett, pp. 292–5; Kurt Lang, "The Military Putsch in a Developed Culture: Confrontations of Military and Civil Power in Germany and France," ed. Jacques van Doorn, *Armed Forces and Society: Sociological Essays* (The Hague, 1968), pp. 203–5.

Austria and the Sudetenland in 1938, and then established a protectorate over Bohemia and Moravia and took Memel in 1939. The army lost confidence in its own judgment. Hitler always seemed to be right and they always wrong. It is hardly to be wondered at that, in the end, the generals followed Hitler blindly into the war, and that they were dazzled by its early successes in Poland and in France. The fact that they had not anticipated these successes either further strengthened Hitler's hand.[40]

Generally it can be said that much of Hitler's power arose out of his willingness to make commitments beyond those of any of his opponents at home or abroad. He was willing to shoulder risks that they would not. Knowing that they were unwilling to risk a major war, he exploited their fears to his advantage till 1939, when the challenge was no more to be avoided. It seemed, to his opponents, that Hitler's commitment approached the total. In the end it exceeded his resources. His technique was to challenge his opponents to raise the stakes secure in the belief that they would, in the end, back down. Yet, as Edward Peterson points out, "Hitler was never as sure of himself and his position as both his enemies and his friends thought. He was a man playing a game of chance, a German roulette, that most men would not join if they could and in which the stakes shook even Hitler." [41] Hitler, it appeared, seemed ready to commit not only the future of his regime but also the future of Germany and the German people, while the German generals were hardly willing to risk even their careers.

Discussing the instances in which the German officer corps lost power would serve little purpose. Even as it stood, an as yet undefeated army, it was increasingly impotent to decide its own fate, let alone that of Germany. Even when defeat stared them unmistakably in the face, they were unable to unite against Hitler and the commitment of the few men of courage willing to do so in 1944 was futile from the start. The German army was decaying at the time of victory and, facing defeat, was already in an advanced state of institutional decomposition.

V Conclusions

Two points emerge: first, that political decay can be said to stem from an unwillingness to commit resources to preserve value positions vital to institutional survival; and, second, that to make commitments involves organizational coherence. Huntington, in the article referred to earlier, cites coherence together with adaptability, autonomy, and complexity as elements in determining level of institutionalization. These have entered indirectly into the discussion for, in the last resort, they help determine the possible level of commitment and are determined in turn by the previous commitments made.

40. Demeter, pp. 199–201.
41. Peterson, *Limits,* p. 431.

Failure to make commitments, to stake resources for value positions, reduces coherence, complexity, autonomy, and adaptability, and hence the level of institutionalization from one time to another. Hence the arguments adduced bear strongly on Huntington's thesis. One of the main points of difference, however, is in the assessment of the role of disorder and institutional decay. There does not seem so close a relationship there. The cases indicate that internally disorganized bodies, at the time of failure, may be in a position of great power, and be high in all institutional indices *related to that power even though unrelated to original function*. This was the case with the French army which was disintegrating *as an army* while gaining in political power. The converse can also be true, for organizations apparently functioning smoothly may, in fact, be in an advanced stage of institutional decay. The German army, still apparently a world-beater in 1942, has indicated this. Indeed the notion that power and function are unrelated, and may be asymmetrical, is one that would bear further research.

Decay, it would seem, first sets in when those steering an institution or organization are afraid to commit resources in defense of value positions vital to its survival. The onset of such decay could well pass unnoticed for a long time as its impact is softened by custom, usage, and the rituals of the organization. The crunch comes when the organization faces a crisis and no resources remain to it. The organization then breaks apart. It is for this reason perhaps that decayed institutions fall so quickly and so little remains after their fall. It took little more than a push to topple the Third French Republic, czarism in Russia, or Nkrumah in Ghana, strong as these had appeared before the testing time. When crisis came, those in charge found they could no longer take decisions with authority, for to make their decisions effective they would have needed resources that either were no longer available, or if available were unutilized because the habit of commitment had been lost.

The distinction made earlier between power and commitment should, perhaps, be again brought back to mind. Loss of power means loss of resources. In the words of Karl Deutsch it means "a loss of resources and facilities required to make the behavior of the system prevail over obstacles in the environment." [42] A comparison can be made to business investment, though this does not hold entirely: a loss of commitment is measured by the failure to invest resources (money, materials, and men). If resources are well invested in business terms, they bring rewards in the form of profits which can be ploughed back in part to increase the money, materials, and men available for further commitment. If the commitment or investment had been a poor one, however, the same resources might have been wholly or partly lost. The element of risk is an obvious one. Investments will, thus, only tend to be

42. Deutsch, *Nerves*, p. 223.

made when on a subjective evaluation of probabilities the magnitude of likely profit outweighs the risk of loss.

There is a similar calculus of risk in politics. Every political decision calls for investment of resources, and miscalculation always means a measure of power is lost. Hence, to take risk involves an act of conscious will. The analogy of politics to business does break down at a vital point: business gains are objectively measurable in terms of money flows, flows of orders, flows of goods and materials, and so on. Political gains are largely subjective, not even votes being a reliable indicator of power. Business resources, being objective, can be hoarded for later use if not used. Political resources, being subjective, tend to be lost if not used. Power in politics, therefore involves a greater need for commitment of resources if the institution is not to decay.

The cases outlined have, it is hoped, made the relationship of political power and the commitment of resources clear. They have shown, perhaps, the intimate relationship between the idea of political power and commitment. The cases have shown this on a relatively small stage, that of the personal interaction of political and military elites. The argument can be broadened with further research. The importance of images of commitment has been realized and developed in the study of international relations.[43] It can be applied also to civil-military relations in a comparative context, and indeed to the comparative study of politics in general.

43. See, for instance, Robert Jervis, *The Logic of Images in International Relations* (Princeton, 1970).

PART VII

Military Man

A DEVELOPMENT MODEL
OF MILITARY MAN*

David Krieger

In this essay I shall refer to a type of individual I have called Military Man
an individual *who supports policies designed to increase his nation's military
preparedness.* He need not actually fight (that is, be a soldier), nor need he
necessarily be bellicose in his interpersonal relations. The orientation of Military Man is simply the advocacy of increased national military preparedness.

Another type of militarily oriented individual is the revolutionary who
seeks to change the social or political order by violent means. The revolutionary is distinguished from the Military Man of this essay by the absence
of his support for his nation's military preparedness; rather than support
the national military, he seeks to defeat it or convert it.

Advocacy of increased military preparedness may be expressed in many
ways. Issues on which expression is possible include increased military
expenditures, increased troop levels and commitments, and increased development and/or deployment of strategic weapons. Expressions of support
may take the form of opinions or attitudes, votes for certain candidates
or issues, or active campaigning for a particular policy.

It is best to consider Military Man as located at one end of a continuum. At the opposite pole from Military Man would be an individual who favors national disarmament and *opposes* increased expendi-

* Reprinted with permission from the *Journal of Contemporary Revolutions,* 1971,
3(1), 68–74.

tures for national military preparedness or military activities. The individual at this end of the continuum may be motivated in his opposition to the national military establishment by non-violent principles, by a revolutionary desire to see the nation weakened, or simply by a pragmatic feeling that the military establishment has grown dangerously large or expensive. An analysis of the reasons for opposing increases in national military power must take into account these diverse motivations. In this article I will concentrate my analysis on the type of individual at the other extreme of the continuum, the Military Man.

It is important to recognize that most individuals lie somewhere between the two poles of strong advocacy of or opposition to increasing national military preparedness. The greatest proportion of citizens of any society are generally willing to accept the policies advocated by their government with neither strong support for or opposition to alterations in national military strength. While the acceptance of the government position may simply be the result of apathy rather than policy commitment, a large deviation from a unimodal distribution of support for the government on this issue may be a preliminary signal of either a governmental shift or strong opposition from either the Right or the Left.

The tendency to accept the governmental position on military strength has a built-in bias toward military development since it generally does not benefit the government to reduce the military establishment unless there are significant public pressures urging this course of action. The prototype of Military Man discussed in this paper lies beyond the modal position of passive acceptance of government policy on this matter. Military Man is seen as an outspoken advocate, or supporter of an outspoken advocate, of increased national military preparedness.

When a government proposes major increases in military expenditures for the deployment of new strategic weapons such as the Nixon administration position on the ABM system, those who acquiesce to government initiatives in these matters may appear quite similar to Military Man. The difference between the acquiescer and the Military Man, however, is that the former would presumably acquiesce to government proposals to reduce military expenditures or disarm while the latter would continue to advocate increased national military preparedness even in light of government initiatives toward the opposite direction. In fact, Military Man might attempt to lead a revolution from the Right if he felt that the government were unresponsive to national security interests.

The developmental model of Military Man outlined below includes the conception of this individual's position on military preparedness as being functional in the satisfaction of certain internal needs. The suggested model derives in large part from work done on the study of the Authoritarian

Personality by Adorno and his co-workers,[1] and subsequent studies relating authoritarianism to foreign policy orientation.[2]

The Authoritarian Personality has been defined as being submissive to authority and domineering of subordinates, aggressive, tough-minded, cynical, superstitious, intolerant of ambiguity, and cognitively rigid. Ideologically, the authoritarian has been found more likely to be conservative in political and economic orientation, nationalistic, and militaristic. The discussion which follows is an attempt to organize these ideological orientations and stylistic traits into a linkage pattern which is explanatory of a developmental model of Military Man.

The basic outline of the development of Military Man shows an evolution from faulty childhood discipline to low self-esteem to misanthropy to authoritarianism to conservatism, nationalism, and militarism. This linkage pattern connecting basic orientations toward the self and mankind with more specific social orientations toward institutional organization, one's nation, and the military is shown in figure 1. The dark arrows indicate

FIGURE I

A Developmental Model of Military Man

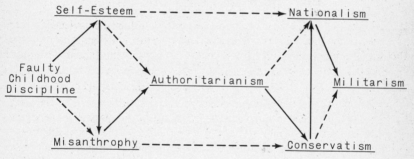

primary linkages, while the dotted arrows represent weaker secondary linkages. The central proposition in this model is that militarism of the variety manifested by Military Man is an adjustment to personal insecurity or low self-esteem brought about by faulty childhood discipline.

Faulty discipline is generally characterized by inconsistency and a lack of warmth in administration. Inconsistency decreases the child's ability to predict the outcomes of his behavior and thus increases his personal insecurity. A lack of warmth or an aloofness on the part of the parents leads the

1. T. W. Adorno et al., *The Authoritarian Personality* (New York: Harper, 1950).
2. William Eckhardt, and Theo Lentz, "Factors of War/Peace Attitudes," *Peace Research Reviews*. I(5) (1967): 1–114.

child to doubt his own value or worthiness. The result of this parental orientation is to make the child feel insecure and unloved. These feelings in the
child may be intensified by the harsh application of physical punishment
often associated with authoritarian childrearing patterns, but, even without
harsh physical punishment, such feelings are devastating to the development
of ego strength in the child.

Obviously a child who perceives a lack of warmth toward him or interest
in him from his parents would feel uncomfortable, at the least. To cope
with this unpleasant situation, the child may attempt to gain esteem outside of home. Peer group gangs, the classroom, or athletic contests may
provide contexts in which a child may gain sufficient esteem to compensate
for the insecurity of the home, but probably not on a permanent basis.
It is likely that the child who feels unloved by his parents will need to be
constantly "proving himself" as self-reassurance that he is a worthy person.
Thus, excellence outside the house and in later life may serve as temporary
insulation against feelings of unworthiness which may reoccur whenever
the "proving" is not sufficiently rewarded.

The "proving" strategy generally is both physically and psychologically
demanding. It requires the individual to structure his environment as a
competition in which he is constantly engaged. This competitive orientation
increases tensions and places him under added stress, which, if raised beyond an optimum level, may decrease performance and reduce the chances
of favorable outcomes. While high tension levels may be preferable to
failure, which forces the individual to confront his feelings of unworthiness,
the tension itself may unfortunately serve to increase the probability of
failure.

Rather than endure the stresses and uncertainties of a "proving" strategy,
the individual may opt for a strategy of judging men in general as evil or
unworthy. A misanthropic view functions to make the self appear less
unworthy "by contrast" with the unworthy state of human nature. If mankind is generally bad, one may reason, then one's own unworthiness is simply
a reflection of the human condition. And individual pays far fewer psychic
costs for the adoption of this strategy, and, if humans tend to orient themselves toward positions of least stress compatible with their goals, then this
strategy has much to recommend it over "proving." It is also likely that,
if faulty discipline results from authoritarian parents, the parents themselves will have adopted the strategy of misanthropy and will model it for
the child.

If one accepts the misanthropic premise that men are generally bad, it
follows that social controls are needed to protect men from each other.
An authoritarian orientation places major emphasis on the hierarchical
ordering of human affairs. When one is a rung up on the ladder, it is his
duty to control those beneath him. At the same time, the authoritarian recognizes the need to submit to the control of those in higher positions in the

hierarchical order. His willingness to take orders makes the authoritarian individual a potentially good Organization Man. If he is tactful enough to be able to give orders without alienating his subordinates, the authoritarian's chances for success as an organizational administrator are high.

The authoritarian reflects his support for the hierarchical ordering of interpersonal relations by favoring social institutions which structure human interactions on this model. Central to the conservative orientation of the authoritarian individual is his support for institutions of social control such as the military and the police.

Domestic conservatism of the authoritarian is expressed in support for expanded police powers and slogans of "law and order." He is likely to favor harsh punishments for those who violate social norms. Believing that deviants must be made examples of so that others will not follow suit, he will argue that sexual offenders should be publicly whipped if not hung and that capital punishment is an appropriate retribution for felons. He will look with a suspicious eye at "radical-liberals" who seem to be espousing greater freedom for mankind and thus, at least potentially, a state of violent anarchy. The reassurance of the conservative comes from "ordered violence" rather than the chaotic variety. If the conservative is upset by "radical-liberals," it may be presumed that he is out-and-out frightened by self-styled revolutionaries such as the Black Panthers, the Weathermen, or even the Yippies.

For Military Man, the nation is a highly prized symbol as long as the government remains powerful enough to enforce "law and order." Should the government become weakened by "infiltration of liberals," Military Man might be persuaded to adopt a revolutionary posture in order to restore the power of the nation. He sees the nation as the primary focus of loyalty, and a symbol of strength. In part, identification with the nation helps to relieve feelings of personal unworthiness. Support of the nation is seen primarily, however, as instrumental in preservation of social order which controls the otherwise nonsocial impulses of unworthy men.

It is a combination of authoritarian, conservative, and nationalistic orientations which shape the international outlook of Military Man. He correctly perceives the international environment to be lacking in order and thus dangerous. He also sees his nation as protecting him from the consequences of international disorder. He recognizes enemies who have the will and, if not guarded against, the capability to destroy his nation (and therefore his strength).

Whereas national enemies are generally not highly salient for most individuals except in time of crisis, for Military Man the enemy is both salient and central in his image of the international environment. In the American context, Military Man would most certainly be a staunch anti-Communist, just as in the Communist context he would be a staunch anticapitalist. It is to cope with the ever-present threat which he perceives to exist in the international environment that Military Man advocates continuously increasing military preparedness for his nation.

In this analysis of Military Man I have not meant to imply that anyone who perceives threats to exist in the international environment is characterized by personal insecurity, misanthropy, etc. Threat may well be observed because it exists, quite independent of an individual's childhood experience self-esteem, etc. However, while there are other strategies for reducing tension in a threatening situation such as negotiation and compromise (often considered by Military Man as appeasement), the primary and often only acceptable solution to Military Man is to be prepared to trade in the currency of military force. Dependence on military force as a problem-solving technique relates back to the misanthropic premise that negotiation and compromise will only forestall the inevitable resort to arms which is inherent in the social relations of men. While others may applaud the development of world law and national participation in international organizations such as the United Nations, Military Man sees these as costly and "unrealistic" diversions from the crucial task of strengthening national military preparedness.

It may be helpful in understanding Military Man's behavior to understand the functions served by holding opinions. In an important study of personality and opinions, Smith, Bruner, and White suggested that opinions may serve three functions for those who hold them.[3] They labeled these object appraisal, social adjustment, and externalization. Object appraisal is akin to reality testing or attempting to bring one's opinions into closer approximation with reality. Object appraisal is useful to the individual in helping him to assess the best (most "realistic") strategies for achieving his goals, and adapting to his environment.

Social adjustment, the second function of opinions, refers to the structuring of interpersonal relationships. The expression of certain opinions are helpful to the maintenance or improvement of relations with other people. One's opinions are often formed so as to facilitate relations with other individuals in a membership group, or to project an image of association with a particular reference group such as "intellectuals," or "good Americans."

The third function of opinions, externalization, is a technique to relieve unresolved inner problems by adopting a particular orientation toward the environment. Techniques used in externalization include projection of hostility and displacement of aggression.

The kinds of opinions expressed by Military Man about the international environment may have some basis in object appraisal and social adjustment,

3. M. B. Smith, J. S. Bruner, and R. W. White, *Opinions and Personality* (New York: Wiley, 1956).

but they seem especially to serve the function of externalization. That is, the opinions of powerful enemies presenting an extreme threat for which the only remedy is the strengthening of his nation's military preparedness function to inhibit anxieties related to low self-esteem and personal insecurity. It is functional to the personal well-being of Military Man to hold the opinions he does.

Military Man's attempts to gather facts or evaluate information are invariably conducted in such a way as to confirm his predispositions about the need for a stronger military. This is to say that in object appraisal information is selectively accepted, perceived and interpreted so as to reinforce existing beliefs. In addition, in all likelihood Military Man joins social groups with like minded individuals so as to achieve social reinforcement for his beliefs about the world.

While the opinions which Military Man expresses about the need for increased national military power may be functional to him as an individual, the conversion of these opinions into national policy may be very dysfunctional to the nation itself and to the general state of the international environment. Policies of increased armaments require increased expenditures from the public pocket, and these revenues are sorely needed to combat ignorance, hunger, disease, poverty, and pollution. The most important dysfunction of the Military Man orientation, though, would be realized in the event of a nuclear exchange.

To the extent that Military Man is driven primarily by internal needs rather than responding to external deeds, he presents a serious threat to world peace. In addition, in a nuclear age, his efforts to increase national security through increased armament work, in fact, to decrease security by adding fuel to the arms race.

What can be done about Military Man? Probably very little. It is unlikely that the household can be penetrated and authoritarian parents deterred from their inconsistent and aloof child-rearing techniques. Perhaps there is some hope that the classroom environment could be structured so as to overcome rather than reinforce faulty child-rearing techniques of the home. Yet, this seems at the moment an idle dream, removed from the practical politics of schoolboard policy-making. Current teacher/student ratios create the conditions for an authoritarian structure in the classroom, and reduce the possibility that a teacher could provide the child from an authoritarian home with the continuity and consistency of warmth not available to him in the home.

Concerted efforts must certainly be made to keep Military Men away from positions of national power where their opinions may exert a strong influence on national policy. How this is done remains at present a question of individual conscience and commitment. Yet, when it becomes clear that the type of individual I have described as Military Man is exerting a dominant influence on society, then it would seem apparent that for the good of mankind generally, organization for change should proceed. Nonviolent techniques may be sufficient in unseating Military Man from power if his ascension to power is recognized early enough and resistance channeled into an effective organizational framework.

CROSS-CULTURAL MILITARISM:
A TEST OF KRIEGER'S
DEVELOPMENTAL MODEL
OF MILITARY MAN*

William Eckhardt

According to Krieger's proposed developmental model of military man (in its most simplified form), faulty childhood discipline causes low self-esteem, which causes misanthropy, which causes authoritarianism, which causes conservatism, which causes nationalism, which causes militarism. There is some evidence to support this model in some of the developed nations of the Western world.

Militarism scales have been validated by the "known groups" method: reserve officers and ROTC students have obtained militarism scores significantly higher than average, while members of peace churches (such as Quakers) and peace groups (including peace research groups) have obtained militarism scores significantly lower than average (Porter, 1926; Eckhardt, Manning, Morgan, Subotnik, & Tinker, 1967). The militarism scale of Eckhardt et al. (1967) was also validated by its predictability of student responses to the Cuban crisis; that is, those students who obtained higher scores on this militarism scale were more inclined to recommend invasion of Cuba during the crisis. The military deterrence factor of this scale was correlated .60 with authoritarianism (Manning, 1964). The total score on this scale was correlated .25 with conservatism, .34 with patriotism, and .35

* This is a slightly revised and updated version of an article which was originally published by the *Journal of Contemporary Revolutions* 3, (2), no. 2 (1971), 113–39. This study was partially supported by grants from Canada Council and Leitch Transport. Computer programming was done by Richard Greenfield and Christopher Young.

with anti-intellectualism (Chesler & Schmuck, 1964); .35 with religious orthodoxy, .39 with anti-Communism, .42 with capitalism, .43 with conservatism, .46 with antidemocratic attitudes, .55 with antiwelfarism, .58 with Goldwaterism, and .66 with aggressive nationalism (Eckhardt et al., 1967). For illustrative purposes, the items in this militarism scale are given in Appendix A of this paper, and illustrative items from some of the other scales mentioned in this and succeeding paragraphs are given in Appendix B. Eckhardt et al. (1967) concluded from their studies, "Militarism in our culture today would seem to be in defense of laissez-faire capitalism, religious orthodoxy, and nationalism, but it would not seem to be in defense of democracy or idealism. On the contrary, it would seem to be on the side of authoritarianism of the right and materialism" (p. 536).

Authoritarianism has been found to be related to militarism and isolationism (Lane, 1955), military ideology (French & Ernest, 1955), aggressive anti-Communism (Fensterwald, 1958; Rosenberg, 1965), and jingoism has been defined as aggressive foreign policy and exaggerated patriotism (Farris, 1960) in the United States.

Militarism was associated with nationalism and dependence upon experts among three hundred Canadian farmers (Kristjanson, undated, but probably in the early 1960's). When Japanese peace-seekers were compared with war-oriented adults, the latter were found to have a lower occupational level, to be less socially oriented and politically active, to be more authoritarian and manipulative, to have a lower sense of civic duty, and to be more conservative (Kuroda, 1964, 1966).

Canadian peace research supporters were relatively high in their economic and educational levels, internationally responsible, welfare-minded, non-dogmatic, and noncynical (Laulicht & Alcock, 1966). On the other hand, the most active attenders of the most dogmatic (Catholic and Fundamentalist) churches in Canada were most in favor of nuclear weapons, least afraid of the nth country problem, wanted bigger and better defense forces, and were most distrustful of peaceful coexistence (Fraser, 1965). On the whole, Canadian Christians were more warlike, and showed no greater sense of responsibility or love, than non-Christians (Alcock, 1965).

These studies have established significant relations between authoritarianism, conservatism, nationalism, and militarism (the last half of Krieger's model). Since many other studies have also shown that religiosity is associated with this ideological complex, religiosity should be included in this half of the model (Adorno, Frenkel-Brunswik, Levinson, & Sanford, 1950; Barton, 1963; Ekman, 1963; Rose, 1963). One such study concludes, "Those who hold Christian religious attitudes strongly are more warlike, less democratic, more punitive, less tolerant, more conservative, less world-minded, more repressive, and less humanitarian than non-Christians" (Russell, 1971, p. 39). As to the personality side (the first half) of Krieger's model, supporting evidence in Adorno et al. (1950), among others, also shows relationships between faulty childhood discipline, low self-esteem, misanthropy, and authoritarianism. Many relations have also been found between personality and ideology (the first and second halves of Krieger's model). For example, people who favored military intervention were shown to feel belligerence under threat (misanthropy), and, according to Gladstone, associated with a tendency to feel threatened (low self-esteem), as found by Gladstone (1955). And high scorers on militarism and nationalism scales gave more aggressive (misanthropic) responses in simulated conflict situations (Crow & Noel, 1955). An average correlation of .28 was found between competitiveness in personal relations (misanthropy) and competitiveness in foreign policy (militarism) for 218 students (Scott, 1960).

Nationalism was significantly correlated .77 with ethnocentrism, .65 with traditional family ideology (strict discipline), .60 with authoritarianism, .52 with religious conventionalism, and .34 with conservatism (Levinson, 1957). Rosenberg's (1956, 1957) misanthropy scale was correlated with authoritarianism and militarism. Over-controllers (presumably subjected to strict discipline) selected military men as great leaders, responded to prestige suggestion, highly valued maintenance of the status quo, referred to power and prestige as criteria for greatness, and chose conservatives as heroes (Blum, 1958). Authoritarianism was loaded .66 to .70 on a rigidity factor (strict discipline), .37 to .48 on a conservative factor, and .27 to .38 on an anxiety or neuroticism (low self-esteem) factor (Rokeach, 1960, p. 420). Among 167 Norwegian military and naval cadets, those who made higher scores on a nationalism scale were likely to be aggressive under threat (Christiansen, 1959). Threat-orientation among these Norwegian subjects was correlated at the interpersonal and international levels.

At the opposite pole of authoritarianism, democratic subjects were better adjusted socially and emotionally (higher self-esteem), and democratic attitudes were highly correlated with tolerant (less misanthropic) attitudes (Lentz, 1943). Deutsch (1960) found that authoritarianism was correlated with a tendency to be suspicious and untrustworthy in a two-person, non-zero-sum game. Children in more authoritarian cultures showed significantly fewer responses of honesty and responsibility (Anderson & Anderson, 1962).

In his survey of international images, Scott (1965) found that those people who had benign images of other nations and who were willing to cooperate with them, were also well informed and personally secure, satisfied, optimistic, and effective (thus demonstrating high self-esteem). On the other hand, people with low self-esteem (poorly informed, dissatisfied with their jobs, lacking a sense of personal worth, pessimistic, and lacking a sense of meaning and purpose in their lives) had threatening images of other nations and felt competitive, ethnocentric, authoritarian, aggressive, and paranoid in relation to them (Rokeach & Eglash, 1956; MacKinnon & Centers, 1956,

1958; Levinson, 1957; Farris, 1960). A sense of individual responsibility for peace was loaded .41 on a factor of mental health, or high self-esteem (Eckhardt et al, 1967).

Students who were more ethnocentric and nationalistic were also more sensitive, schizoid, depressive, and antisocial (Spilka & Struening, 1956). Rokeach's (1960) factor analyses showed that neuroticism (low self-esteem) was loaded with anxiety, paranoia, self-rejection, authoritarianism, opinionation, and rigidity. In another study, anxiety (associated with low self-esteem) was claimed to be at the root of rigid foreign policy, and correlated highly with both a negative image of Russia, and poor knowledge of foreign affairs (Modigliani, 1961). Religiously devout subjects were less humanitarian, more punitive, less tolerant, and more anxious (Rokeach, 1965).

What these studies conclude is that neuroticism (low self-esteem) and misanthropy (or hostility) contribute to authoritarianism, dogmatism, militarism, nationalism, and religiosity. Studies reviewed elsewhere (Eckhardt, 1971a) show that neuroticism also contributes to conservatism, so that the ideological variables in the second half of Krieger's model may be at least partly interpreted as ego-defense mechanisms, that is, as ways of compensating for low self-esteem associated with faulty childhood disciplines, thus validating the whole model to some extent. However, although most studies have shown some relationship between ideology and personality, Comrey (1966) has found no such relationship. Consequently, the model may be expected to hold as a general rule, but hardly as a universal law. Furthermore, this general rule has found support so far primarily from studies conducted in six of the developed nations of the Western world: Australia, Canada, England, Japan, Norway, and the United States. The question remains as to whether this model can be generalized to apply as well to other parts of the world.

During the last few years, a number of Canadian studies have provided rather detailed confirmation of Krieger's model (Eckhardt & Newcombe, 1969; Eckhardt, 1969a, b, c; Eckhardt & Alcock, 1970; Eckhardt, 1971b). These studies have shown militarism to be related to authoritarianism, conservatism, nationalism, and religiosity on the ideological side, and to faulty childhood disciplines, misanthropy, and neuroticism (low self-esteem) on the personality side. These studies, however, like the other studies reviewed in this section, have been limited to subjects in the Western world. The purpose of the present paper is to find how far these results can be generalized to the East as well as to the West, the East including poorer and non-white nations in contrast to the richer white nations of the West.

Methods and Subjects

A multi-national student survey (MSS) was designed by Finlay, Iversen, & Raser (1969), which was administered to about 5000 male university students (ages 22 to 24) in 18 nations by 42 researchers in 1968 and 1969. The names of these researchers, the characteristics of their institutions, and the complete questionnaire is available in Finlay, Iversen, & Raser (1969). The 18 nations included nine developed Western nations in North America and Europe, six Eastern nations in Africa and Asia, and three miscellaneous nations (Brazil, an unidentified Communist country, and South African whites).

The MSS questionnaire included about 200 items, seven of which were similar to items usually found in a militarism scale. The present study was based on the correlations between one of these items and 93 other variables, including opinion items and personality scales. The item selected to represent militarism in this study was worded as follows: *"My country should (not) start to disarm."* This item was chosen because the sum of its correlations with other variables was generally higher than that of any of the other militarism items. This criterion permitted as many significant correlations as possible to emerge, which partially compensated for the low reliability of items. This in no way biased the results, since the correlations had as much opportunity to negate the Western model of militarism as to confirm it.

Results

Table 1 shows the names of the nations and the number of students in each nation who completed the item of militarism vs. disarmament. Table 1 also provides the mean score and standard deviation of this item for each national sample and group of samples. Since this item was rated by the students on a 7-point scale, a mean score of 4 would be the mid-point where the students would be equally divided for and against disarmament. Since the average mean score of the 18 national samples was 4.59, as shown in the last row of Table 1, the majority of these students (60% of them) were opposed to disarmament.

Table 1 shows that Danish, Communist, Dutch, German, Ceylonese, and Canadian students obtained the lowest mean scores on this item of militarism, while white South African, Nigerian, Finnish, South Korean, Australian, Brazilian, and Indian students obtained the highest mean scores. The Eastern and miscellaneous means were significantly higher than the Western mean on this item. While 60% of these students were militaristic on this one item, 45% of them were militaristic on the average of all seven militaristic items. South African whites (60%), Indians (58%), Nigerians (53%), Australians (52%), and United States (51%) were significantly higher than average on all seven items, while Danes (35%), Germans (36%), and Communists (36%) were significantly lower than average. Eastern students (50%) were significantly higher than Western students (42%). Among Western students, the American group of nations (49%) was significantly higher than the European (39%) and Scandinavian (38%) groups, and "American" stu-

Table 1: *Mean Militarism Scores and Number of Significant Correlations*

Student Samples	Number	Average Militarism	Standard Deviation	Number Significant r's	Percent Predicted Direction
Australia (Au)	140	5.31	1.91	46	100%
Canada (Ca)	272	4.03	2.02	64	100%
United States (US)	650	4.33	2.06	68	94%
America (Am)	*1062*	*4.39*	*2.07*	*74*	*97%*
England (En)	60	4.05	2.16	33	94%
Netherlands (Ne)	914	3.56	2.20	83	92%
West Germany (WG)	65	3.09	2.11	45	98%
Europe (Eu)	*1039*	*3.56*	*2.20*	*82*	*94%*
Denmark (De)	304	3.01	2.26	69	94%
Finland (Fi)	269	5.71	1.84	61	88%
Sweden (Sw)	196	4.26	2.20	57	96%
Scandinavia (Sc)	*768*	*4.29*	*2.41*	*71*	*94%*
Brazil (Br)	170	5.24	2.12	17	82%
Communist (Co)	289	3.35	2.07	31	58%
South Africa White (SAW)	101	6.69	0.81	16	88%
Miscellaneous (Misc)	*560*	*5.09*	*1.67*	*NA*	*NA*
Ghana (Gh)	293	4.66	2.13	24	75%
Nigeria (Ni)	294	6.13	1.67	19	74%
South Africa Black (SAB)	149	4.59	2.11	15	87%
Africa (Af)	*736*	*5.23*	*2.09*	*40*	*73%*
Ceylon (Ce)	236	4.01	2.23	24	67%
India (In)	278	5.20	2.14	42	67%
South Korea (SK)	299	5.55	1.66	25	60%
Asia (As)	*812*	*4.99*	*2.07*	*48*	*67%*
East (N = 6)	1548	5.10	2.09	52	75%
West (N = 9)	2869	4.08	2.23	NA	NA
Misc (N = 3)	560	5.09	1.67	NA	NA
Total (N = 18)	*4977*	*4.59*	*2.08*	*NA*	*NA*

Notes.—Most of these averages and standard deviations were computed from the scores of complete samples, except that Miscellaneous, West, and Total scores were obtained simply by averaging the national means and standard deviations within these groups. The abbreviation for each nation and for each group of nations is given in parentheses in this table, to serve as a reference for their use in the following tables.

dents were not significantly different from Eastern students on their mean score for all seven items.

Table 1 shows the number of significant correlations found between militarism vs. disarmament and the 93 other variables included in this study.

The majority of these correlations were in the predicted direction (that is, they were consistent with the findings reviewed in the introduction to this paper) for every national sample, as shown in the last column of Table 1, suggesting that the concept of militarism as it has been defined by studies in some Western developed nations may be generalized to some extent to all of the nations included in this study, the generalization being least reliable for Communist and South Korean students. However, it should be noted that the number of significant correlations varied from as low as 15 for South African blacks (which was at least partly attributable to the small number of these subjects) to as high as 83 for Dutch students (which was at least partly attributable to the large number of these subjects). Since the number of subjects varied from nation to nation, the number of significant correlations were not directly comparable, since smaller samples required a higher correlation in order to achieve significance.

Table 2 shows those variables which were generalizable from West to East, and which were not. These variables were divided into four general categories for convenience of presentation and discussion. The *affective* category included five of Gordon's (1960) personality scales, five of Cantril's (1965) optimism items, and one item affirming the value of the family. The *behavioral* category included those items referring to individual activities and social affiliations. The cognitive category included international knowledge as self-rated and as measured by responses to an objective test of seven items. The *ideological* category included opinions, interests, and values concerning political, religious, and social affairs in general.

The criterion used for selecting the variables in Table 2 was that each variable should be significantly correlated with militarism vs. disarmament in the predicted direction for at least one third of the Western nations in this sample. The actual number of nations which contributed to this criterion for each variable is shown in Table 3.

An asterisk before any variable in Tables 2 and 3 indicates that that variable was positively correlated with militarism in at least some part of the East as well as in the West. A minus sign before any variable in Tables 2 and 3 indicates that that variable was negatively correlated with militarism in the East, while it was positively correlated with Western militarism. A fraction before any variable indicates that proportion of the variable which was significantly correlated with militarism in the East as well as in the West. The number in parentheses following the name of each variable indicates the number of items included in that variable. Only significant correlations (based on a one-tailed test) have been entered in Table 2, so that blank spaces indicate no significant correlation. The correlations in Table 2 were between items only, except for the personality scales and the international knowledge scale. When there was more than one item per variable, the correlations were obtained simply by averaging the coefficients across all items in any variable. Consequently, almost all of the correlations in Table

Table 2: *Correlations with Militarism vs. Disarmament*

	Am	Eu	Sc	Br	Co	SAW	Af	As	East	West
Affective										
* No Personal Benevolence (15)	23	27	30				9	13	10	27
— Personal Conformity (15)	18	25	42	18	13		−7		−4	28
— Family Value (1)	19	17	41			18	−13			26
— National Optimism (3)	22	19		18				−6		13
* Personal Leadership (16)	14	13	14				10	13	11	14
** No Personal Support (15)		10	18					10	5	11
** Personal Optimism (2)		9						8		4
— No Personal Independence (16)	10		7					−12	−4	7
Behavioral										
— No Demonstrations (1)	24	32	28							28
— No Intl. Discussion (1)	16	14		−16						11
— Natural Sciences (1)	16	19	28							21
* Military Service (2)	11	19	26				8		4	19
— Religious Affiliation (1)	13	5	34							17
* Socio-Economic Status (2)		13		−14	−13			14	5	3
— No Newspaper Writing (1)	8	6					−11			4
— No public Speaking (1)	5			−10					−8	
Cognitive										
— Less Intl. Knowledge (Self)		6	9						−4	5
— Less Intl. Knowledge (7)		8	14				−7		−4	8
Ideological										
* Militarism (6)	32	35	30	14			13	12	12	32
* Western Orientation (2)	24	40	44					19	10	36
** Conservatism (Self-Rated)	38	54	43			16		13	9	45
½ Hereditary Theories (2)	21	29	33							28
** Vs. World Government (3)	26	26	23			20	13		7	25
— Nothing to Learn (1)	26	32	24	13						27

Variable									
* Opposition to Drugs (2)	20	22	33			18	7	4	25
* Private Property (2)	33	41	35			17	10		25
− Racialism (2)	23	27	27				−7	11	25
* Nationalism (3)	21	19	18			11	10	11	19
⅓ Resistance to Social Change (3)	26	38	32	15			7	5	32
* No National Benevolence (2)	14	31	32			8			26
* Competition (1)	27	28	40				7	−6	32
− Religiosity (2)	17	11	34				−8	7	21
* National Leadership (2)	16	15	15				9		15
− Foreign Aid Not Exploitative (1)	21	27	28			−12	7	5	25
* Vs. Internationalism (5)	13	19	17						16
− Technology (2)	18	10	18						15
* No National Recognition (1)	14	−7		11		9	−11	13	16
* Vs. Strong Social Inst. (1)		22	24	−15		7	19	10	5
− National Interest (1)	7	21	−13			7	13	4	6
− National Support (2)	8	7				9			8
− No National Independence (2)		16	9	12		−6	−6		11
− Future Orientation (3)	7		24						11
− No Political Interest (1)	12	15	7	−12					11
− National Conformity (2)		20	12						10
* Wars Real Interest Conflicts (1)	8						9	6	4
* No National Interference (1)	−23	7	26		10			6	3
− No Govt. Population Control (1)	7	8	7					−5	7
− No Central Planning (1)				−12	−23	11		6	3

Notes.—The meaning of the abbreviations of nations and groups of nations at the top of each column is given in Table 1. An asterisk in front of any variable indicates that that variable was significantly correlated with militarism vs. disarmament in Africa, Asia, or the East as well as in America, Europe, Scandinavia, or the West. A fraction in front of any variable indicates the proportion of that variable which was significantly correlated with militarism in the East as well as in the West. A minus sign in front of any variable indicates that that variable was negatively correlated with militarism in the East.

The number in parentheses following each variable indicates the number of items included in that variable. With the exception of Gordon's (1960) personality scales under the affective category and the international knowledge scale under the cognitive category, all correlation coefficients in the body of the table were simply averages of the correlations of each variable's items with the militarism vs. disarmament item. Blank spaces indicate no significant correlations.

Table 3: *Number of Nations with Variables Significantly Correlated with Militarism vs. Disarmament*

	West (N = 9)	East (N = 6)	Misc (N = 3)	Total
Affective				
* No Personal Benevolence (15)	9	3	0	12
Personal Conformity (15)	9	1-1 (Ce)	2 (Br, Co)	12-1
Family Value (1)	8	1-1 (Gh)	1 (SAW)	10-1
National Optimism (3)	7-1 (Fi)	0	2 (Br, Co)	9-1
Personal Leadership (16)	4	1 (Ni)	0	5
No Personal Support (15)	4	1 (In)	0	5
Personal Optimism (2)	4	0	1 (Br)	5
— No Personal Independence (16)	3	1-3	0	4-3
Affective Average (83)	*6*	*1-1*	*1*	*8-1*
Behavioral				
No Public Demonstrations (1)	9	1-1 (SK)	0	10-1
* No Intl. Discussion (1)	6	3	0-2 (Br, Co)	9-2
Natural Sciences (1)	6	1 (Ce)	0	7
Military Service (2)	7	0-1 (In)	0	7-1
Religious Affiliation (1)	5	0	0	5
* Socio-Economic Status (2)	3-½ (Ca)	2-½ (SK)	0-2 (Br, Co)	5-3
No Newspaper Writing (1)	3	0-1 (Ni)	0	3-1
No Public Speaking (1)	3-1 (Ne)	0-1 (Ni)	0	3-2
Behavioral Average (10)	*5*	*1-½*	*0-½*	*6-1*
Cognitive				
Less Intl. Knowledge (Self)	4	0	0-1 (Co)	4-1
Less Intl. Knowledge (7)	3	0-1 (In)	0	3-1
Cognitive Average (8)	*3½*	*0-½*	*0-½*	*3½-1*
Ideological				
* Militarism (6)	8	3	1 (Br)	12
** Western Orientation (2)	9	3	0	12
* Conservatism (Self-Rated)	9	2 (Gh, SK)	1 (SAW)	12

Variable				
Hereditary Theories (2)	9	1	1 (Co)	11
* Vs. World Government (3)	8	3–1 (As)	0	11–1
— Nothing to Learn (1)	9	1–2 (Gh, SK)	1 (Br)	11–1
* Opposition to Drugs (2)	7	2 (Ni, As)	1 (SAW)	10
* Private Property (2)	8	2 (As)	0	10
Racialism (2)	9	0–1 (As)	1 (SAW)	10–1
* Nationalism (3)	6–1 (Sc)	3	1 (Br)	10–1
⅓ Resistance to Social Change (3)	8	1–1 (As)	1 (SAW)	10–1
** No National Benevolence (2)	7	2 (Gh, In)	0	9
Competition (1)	8	1 (In)	0	9
— Religiosity (2)	7	1–2 (As)	1 (Co)	9–2
National Leadership (2)	6	1 (Gh)	1	8
Foreign Aid Unexploitative (1)	8	0	0	8
Vs. Internationalism (5)	6	1–1 (As)	1	8–1
* Technology (2)	5	2–1 (As)	1 (Br)	8–1
** No National Recognition (1)	4–2 (Ne, De)	2 (Ni, Ce)	1 (Co)	7–2
* Vs. Strong Social Inst. (1)	4 (Ne, Sc)	2–1 (SAB)	0	6–1
National Interest (1)	5	1 (Ni)	0–1 (Co)	6–1
National Support (2)	5–1 (WG)	1–1 (In)	0–1 (SAW)	6–3
Future Orientation (3)	3	1	1	5
No Political Interest (1)		1 (Ce)	0–1 (Co)	5–1
No National Independence (2)	4–1 (Fi)	0–2 (Ce, In)	1 (Co)	5–3
National Conformity (2)	3–1½ (US)	1½–1 (Ni)	0	4½–1½
Wars Real Interest Conflicts (1)	4	0	0	4
— No National Interference (1)	3–2 (Au, US)	1–1 (SK)	0	4–3
— No Govt. Population Control (1)	2 (Ca, Ne)	0–2 (Gh, In)	0–1 (Co)	2–3
No Central Planning (1)	2 (Ne, De)	1–1 (SK)	0–2 (Co, SAW)	3–3
Ideological Average (60)	6–½	1½–½	½–⅓	8–1

Notes.—An asterisk in front of any variable indicates that that variable was significantly correlated with militarism vs. disarmament in at least one-third of the nations in the East as well as in the West. A fraction in front of any variable indicates that proportion of that variable which was significantly correlated with militarism vs. disarmament in at least one-third of the nations in the East as well as in the West. A minus sign in front of any variable indicates that that variable was negatively correlated with militarism vs. disarmament in the East. The meaning of the nation abbreviations in parentheses in the body of the table may be found in Table 1.

2 were attenuated or reduced in size by the unreliability of the items. When corrected for this attenuation, the true correlations would probably be about twice as large as those reported here. The fact that approximately half of the variables were skewed in the distribution of their responses would constitute another source of attenuation, tending to reduce the size of the correlation coefficients. In addition, an acquiescence or agreement response set was found among African, Asian, and Brazilian students, which would further reduce the size of their observed correlation coefficients, which should probably be multiplied about four times to approximate the true correlations. Consequently, these results provide a very conservative estimate of the similarities and differences between East and West. Finally, it should be noted that the variables under each category in Tables 2 and 3 have been listed in the order of their generalizability, that is, according to the number of nations where they were positively correlated with militarism, as recorded in the last column of Table 3.

Affect

An illustration of each one of the affective variables associated with Western militarism is given here in the order of their appearance in Tables 2 and 3. No personal benevolence: "It is least important to me to make friends with the unfortunate." Personal conformity: "It is most important to me to follow rules and regulations closely." Family value: "The family is necessary to preserve fundamental human values." National optimism: "My greatest hopes for my country are well satisfied." Personal leadership: "It is most important to me to hold an important job or office." No personal support: "It is least important to me to have others agree with me." Personal optimism: "My greatest hopes for myself were well satisfied five years ago." No personal independence: "It is least important to me to be in a position of not having to follow orders."

The asterisks and minus signs under the affective category in Table 2 show that one-half of these variables were positively correlated with militarism in the East as well as in the West, but that the other half of these variables were negatively correlated with militarism in the East although they were positively correlated with militarism in the West. In the East as in the West, militarism was positively correlated with no personal benevolence, personal leadership, no personal support, and personal optimism. However, in the East, unlike the West, militarism was negatively correlated with personal conformity, family value, national optimism, and no personal independence. These findings would suggest that Eastern militarism, so far as affective variables were concerned, was half the same as Western militarism, but half different from it. Since the chief differentiating variables were personal conformity, family value, and national optimism, this finding would suggest that militarism was consistent with personal, familial, and national conformity in the West and, to a lesser extent, in the miscellaneous nations, but that Eastern conformity was opposed to militarism. Consequently, it would be inferred that the meaning of conformity in the East must differ markedly from its meaning in the West. While conformity vs. independence is generally undesirable in the West, this combination of interpersonal values would seem to be generally desirable in the East.

Table 3 shows that the significant correlations found for Africa, Asia, or the East as a whole, were not generally found in the individual Eastern nations, except for no personal benevolence and personal independence.

The first column of Table 4 shows the percentage of affective variables shared by the various nations or groups of nations included in this sample.

Table 4: *Shares in Western Militarism*

Student Samples	Affective	Behavioral	Cognitive	Ideological	Total
America	75%	88%	0%	80%	61%
Europe	88%	88%	100%	90%	92%
Scandinavia	75%	50%	100%	80%	76%
Brazil	25%	−25%	0%	10%	3%
Communist	13%	−25%	−50%	−3%	−16%
South Africa White	13%	0%	0%	10%	6%
Africa	0%	0%	−50%	27%	−6%
Asia	25%	13%	0%	27%	16%
East	13%	13%	−100%	43%	−8%

Notes.—This table was directly derived from Table 2. The percentages in the body of this table simply represent the proportion of variables in each category of Table 2 which were significantly correlated with militarism vs. disarmament for each nation or group of nations. The last column in this table is simply the average of the four categorical percentages.

These percentages were obtained by subtracting the number of negative correlations in Table 2 from the number of positive correlations and dividing by the total number of affective variables. While the groups of developed Western nations shared in at least 75% of these variables, the miscellaneous and Eastern nations shared in only 25% of these variables at the most. Consequently, the affective variables associated with Western militarism cannot be generalized to other nations without some limitations and qualifications.

Behavior

An illustration of each of the behavioral variables associated with Western militarism is given here in the order of their appearance in Tables 2 and 3. No public demonstrations: "I have never participated in a political demonstration." No international discussion: "I almost never discuss the international situation." Natural sciences: "My major area of study is in the natural sciences, medicine, engineering, or law." Religious affiliation: "I have some religious affiliation." Socio-economic status: "My father's or guardian's occupation is (was) executive, professional, managerial, or administrative." No newspaper writing: "I have never written to a newspaper to present a point of view." No public speaking: "I have never spoken at a meeting to present a point of view."

Table 2 shows under the behavioral category that military service and socio-economic status were significantly correlated with militarism in the East as well as in the West, but no public speaking was negatively correlated with Eastern militarism. Table 3 shows that no international discussion and higher socio-economic status were significantly correlated with militarism in the East as well as in the West, but these two variables were negatively correlated with militarism among Brazilian and Communist students. Table 4 shows in the behavioral column that the three groups of Western nations shared at least 50% of these behavioral variables (88% for the American and European nations), while the miscellaneous and Eastern nations shared only 13% of these variables at the most. Consequently, the behavioral variables associated with Western militarism cannot be generalized to other nations without considerable qualifications and limitations.

Cognition

The two cognitive variables associated with Western militarism may be illustrated as follows: Less international knowledge (self-rated): "I am less informed about foreign news compared with other students in my field." Less international knowledge (objectively measured): "Zambia is militarily stronger than Rhodesia."

Table 2 shows under the cognitive category that Eastern militarism was positively correlated with international knowledge, while Western militarism was negatively correlated with international knowledge. Table 3 shows that these relationships held for only one nation in the East (India) and for only three or four nations in the West (Europe and Scandinavia). The cognitive column in Table 4 shows these relations in percentage form. In general, these results would not suggest a very strong relationship between militarism and knowledge, so far as these measures were concerned, but the relationship was negative for Scandinavian and West European students while it was positive for African and Communist students.

Ideology

The first 18 ideological variables associated with Western militarism may be illustrated as follows, in the order of their appearance in Tables 2 and

3. Militarism: "My country should have its own nuclear weapons." Western orientation: "My country has more to learn from the West than from the East." Conservatism (self-rated): "Compared with other students at my university or college, my political position is to the right." Hereditary theories: "War is a result of the inherent nature of men." Vs. world government: "My country should never give up any of its sovereignty to a supranational or world institution." Nothing to learn: "My country has not much to learn from the culture of many of the new nations." Opposition to drugs: "Drugs such as marijuana should not be used even if non-addictive and physically harmless." Private property: "The institution of private property is a sound basis on which to build a society which fulfills the needs of its members." Racialism: "One of my primary loyalties is to my race." Nationalism: "One of my primary loyalties is to my country." Resistance to social change: "The basic organization of our society should not be fundamentally changed." No national benevolence: "It is least important to me for my nation to share its wealth with other poorer nations." Competition: "Competition is an effective way of promoting social progress." Religiosity: "The restraints imposed by strong religious institutions are essential to curb man's natural instincts." National leadership: "It is most important to me for my nation to be a leader in its relationships with other nations." Foreign aid not exploitative: "Giving aid is a technique used by rich nations to help poor ones, but not to exploit them." Vs. internationalism: "The economic improvement of the poor nations is not essential to world peace." Technology: "Technological advances can solve most of the serious problems of the human race."

Of the 30 variables listed in Table 2, 29 of them were associated with Western militarism in at least one group of Western nations. Seventeen of them were positively associated with militarism in the East as well as in the West. Only nine of them were associated with militarism in at least one of the three miscellaneous nations. Seven of these variables were negatively associated with Eastern militarism: Racialism, religiosity, the belief that foreign aid is not exploitative, technology, national support, no national independence, and no government population control. A similar pattern of relations is shown by the asterisks and minus signs in Table 3. Table 4 shows in the ideological column that at least 80% of these ideological variables were shared by the three groups of Western nations, while 43% of them were shared by the Eastern nations at the most, and 10% of them (at the most) were shared by the three miscellaneous nations. Consequently, Eastern militarism was at least half like Western militarism in its ideological beliefs, so that the ideological correlates of Western militarism can be generalized to the East as well, with some qualifications and limitations, but very few of these ideological correlates can be generalized to the miscellaneous nations.

The fact that the item of militarism vs. disarmament was significantly correlated (on the average) with six other militarism items in only 12 of the 18 nations, as shown in Table 3, would suggest that 12 may be the upper possible limit due to the various sources of attenuation mentioned previously.

Discussion

The results of this multi-national analysis of militarism is generally consistent with Krieger's model of military man so far as nine Western developed nations are concerned. It will be recalled that, according to Krieger's model, militarism was linked to the ideological variables of authoritarianism, conservatism, and nationalism, and to the personality variables of low self-esteem, misanthropy, and strict childhood discipline.

Measures of childhood discipline, as such, were not included in the MSS questionnaire, but the affective variables of personal conformity vs. independence, with their emphases on rules and regulations, would suggest the hypothetical inference of strict childhood discipline. Since the personality scales used in the MSS questionnaire were composed of forced-choice items, equated for social desirability, low self-esteem (or neuroticism) was not directly tapped by any of these scales. However, other studies of conformity (Crutchfield, 1955; Barron, 1963, Ch. 14; Elms & Milgram, 1966; Hampden-Turner, 1970, Ch. 5) have shown that conformists were lower in ego strength and self-confidence (both indices of low self-esteem or neuroticism) than were dissenters. Consequently, it can be inferred that the relation between personal conformity and Western militarism implies a relation between low self-esteem and militarism in the West.

Since the lack of personal benevolence in the present study was probably similar to the concept of misanthropy, and since lack of benevolence was related to militarism both East and West, this part of Krieger's model has been confirmed for both the East and the West, but not for the miscellaneous groups of Brazilian, Communist, and South African white students.

The next variable in Krieger's model was authoritarianism. If the desire for personal leadership implies authoritarianism, then authoritarianism was related to militarism among both Eastern and Western students. But if authoritarianism is limited to racial prejudice (racialism in the present study), then authoritarianism was related to militarism among Western students but not among Eastern students.

Conservatism, as self-rated and as indicated by favorable attitudes toward competition and private property, was related to militarism among Asian and Western students, but not among African students.

From the results of this study, it would seem that the ideological part of Krieger's model holds fairly well for military man in the East as well as in the West. Although the personality part of his model seems to fit the West very well, only misanthropy (as operationalized by lack of benevolence) can be generalized to the East as well, but not to the miscellaneous nations.

However, both strict discipline and low self-esteem were inferred to be related to Western militarism because of their relationship to the values of conformity vs. independence. But the combination of conformity vs. independence seems to have a different meaning in East and West, as already noted. Since this combination is generally undesirable in the West, its opposite (independence vs. conformity) might be just as undesirable in the East. While conformity tends to be associated with leadership vs. benevolence among American males, especially infantry officers and management personnel (military-industrial complex), conformity tends to be associated with benevolence vs. leadership among oriental students of both sexes (Gordon, 1967). Consequently, this difference in the meaning of conformity in the East and West, such that Eastern conformity may imply social responsibility while Western conformity implies lack of social responsibility, would lead to the hypothetical inference that independence vs. conformity in the East might be related to strict discipline and low self-esteem even as conformity vs. independence is related to these two variables in the West. This hypothetical inference is strengthened by the fact that Morris (1956) found that Eastern values were more socially concerned while Western values were more self-indulgent, so that conforming to the Eastern culture would be conducive to the development of social responsibility (benevolence) while conforming to the Western culture would mean social irresponsibility (lack of benevolence). If this lack of social responsibility is a function of strict discipline in the East as it is in the West, and if strict discipline in the East lowers self-esteem there as it does in the West, then the present results at least partially support the personality part of Krieger's model as well as the ideological part in the East as well as in the West.

However, the data fit the model only as a general rule and not as a universal law. Not every nation in our sample showed significant relations between militarism and every other part of the model, and the miscellaneous nations showed many exceptions to the general rule. Strictly speaking, Krieger's model does not require militarism to be directly related to every other part of the model, but only to its ideological parts (authoritarianism, conservatism, and nationalism). The ideological half of the model is then joined to the personality half by authoritarianism alone. Consequently, the test of Krieger's model made in this paper has been a very stringent test, to say the least.

Summary

According to Krieger's developmental model of military man, strict discipline causes low self-esteem, which causes misanthropy, which causes authoritarianism, which causes conservatism, which causes nationalism, which causes militarism. Many previous studies in several Western developed nations would confirm the general outlines of this model. The purpose of the present study was to find how well this model fit militarism in the East as well as in the West.

The results of a multi-national student survey were used to test Krieger's model in Western and non-Western nations. An item measuring militarism vs. disarmament was correlated with 93 other variables, including affective, behavioral, cognitive, and ideological items and scales.

At least two-thirds of these variables were significantly correlated with militarism in the direction predicted by previous studies for 16 of the 18 national samples of students, but somewhat less than this for Communist and South Korean students (58% and 60%, respectively). Since these correlation coefficients were attenuated by several sources (unreliability of items, skewedness of half of the response distributions, and acquiescence response set among African, Asian, and Brazilian students), the reported similarities and differences among national samples were conservative estimates of the true similarities and differences.

Since these sources of attenuation were such as to result in the militarism vs. disarmament item's achieving an average significant correlation with six other militarism items in only 12 of the 18 national samples, this number would presumably be the highest number of national samples in which the correlations between militarism vs. disarmament and other variables might be expected to be significant. Since militarism was significantly correlated with nationalism in 10 other nations, this result would constitute a satisfactory confirmation of this link in Krieger's model. Further confirmations would be provided by the relations between militarism and self-rated conservatism in 12 nations (including private property in 10 nations and competition in 9 nations), between militarism and authoritarianism (as operationalized by racialism in 10 nations and by personal leadership in only 5 nations), between militarism and misanthropy (as operationalized by lack of personal benevolence in 12 nations), between militarism and low self-esteem and strict discipline (as inferred from personal conformity) in 12 nations.

The negative relationship between militarism and personal conformity in the East was believed to reflect a basic difference between Eastern and Western cultures, so that conformity in the East was believed to reflect social responsibility while Western conformity was believed to reflect a lack of social responsibility. It would be further hypothesized that other East/West differences follow from this difference in the meaning of conformity. For example, personal independence in the West is generally socially responsible and non-conformist, while personal independence in the East may simply mean "selfishness"; racialism and religiosity in the East may be socially responsible, while in the West they are more often than not associated with lack of responsibility. These East/West differences need much more research before their meanings can be established more definitely than has been possible with the aid of merely hypothetical reasoning.

The general summary of the findings of this study in relation to Krieger's developmental model of military man are presented in Table 5.

If we suspend judgment concerning the exceptions to the first two links in Krieger's model, then the last four links hold fairly well in the East as well as in the West, but the three miscellaneous nations remain outstanding exceptions to the general model, at least so far as its parts have been operationalized in this study. Further research on this model would seem to be well warranted by the finding that it does seem to provide a general rule for the understanding of militarism, although it falls far short of constituting a universal law.

Political Implications of Krieger's Model

Since the factor of militarism which was extracted from two Canadian studies (Eckhardt, 1969c) fit the model of military man proposed by Krieger, the policy implications of those studies would also apply here in general. So far as militarism and its ideological correlates of authoritarianism, conservatism, and nationalism are generally related to faulty personality traits (low self-esteem and misanthropy, indicating lack of trust in oneself and others), militarism and its policy of military deterrence would seem to be more of an ego-defensive mechanism than a reality-testing function. Consequently, there is no reason to believe that militaristic attitudes provide a useful guide to human health and welfare in general, and there is some evidence to suggest that militarism is opposed to welfarism in the minds of men.

Since military ideologies and suspicious, threat-oriented personalities are both functions of faulty childhood disciplines, more freedom from these disciplines in childhood should contribute to the development of more socially responsible adults. Since childhood discipline is but one aspect and expression of the authority structure of a society as a whole, authority structures in general should be changed to become more permissive and responsible to human needs in order to contribute to the development of more freedom and responsibility in human relations, including international relations.

Also, as a matter of policy, more nations should establish peace research institutes so as to facilitate the cross-cultural studies of attitudes, events, and institutions related to war and peace, for the sake of human survival and the quality of life which survives. Those few nations who have already taken this step (the Scandinavian countries, Netherlands, and Germany) deserve the greatest praise for their sense of social and international responsibility. Those affluent nations which have not yet taken this step deserve to be censured for their lack of social and international responsibility. It should be carefully noted that a peace research institute (unlike any number of national intelligence agencies, which serve one nation *against*

Table 5: *Confirmations and Denials of Krieger's*
Developmental Model of Military Man

Causal Chain	Confirmation	Denial
Strict Discipline	Strict discipline hypothetically linked to militarism by way of personal conformity vs. independence among Brazilian, Communist, and Western students.	Personal conformity vs. independence negatively related to militarism among African students. No significant relationship among Asian and South African white students.
Low Self-Esteem	Low self-esteem linked to militarism among Brazilian, Communist, and Western students by way of studies relating conformity to low ego strength and low self-confidence.	Ditto above. But if Eastern conformity means something quite different from Western conformity, these first two denials may turn out to be confirmations.
Misanthropy	Misanthropy linked to militarism among Eastern and Western students by relationship between militarism and lack of personal benevolence (assuming this lack is equivalent to misanthropy).	Lack of personal benevolence not significantly related to militarism among Brazilian, Communist, and South African white students.
Authoritarianism	So far as authoritarianism is implied by personal leadership, then authoritarianism is related to militarism among both Eastern and Western students. So far as authoritarianism is implied by racialism, then authoritarianism is related to militarism among Western and South African white students.	Personal leadership not significantly related to militarism among Brazilian, Communist, and South African white students. Racialism negatively related to militarism among Asian students, and no significant relation among African, Brazilian, and Communist students.
Conservatism	Self-rated conservatism related to militarism among Asian, Western, and South African white students. Competition and private property related to militarism among Asian and Western students.	No significant relationship among African, Brazilian, and Communist students. No significant relationship among African, Brazilian, Communist, and South African white students.
Nationalism	Nationalism related to militarism among Brazilian, Eastern, and Western students.	No significant relationship among Communist and South African white students.

Notes.—The confirmations and denials in this table are based on the correlation coefficients in Table 2.

another) is established by a nation as an autonomous research agency, none
of whose results and publications are restricted, but all of whose projects
are freely chosen and published for the benefit of all nations.

Appendix A. Militarism Scale (Eckhardt et al., 1967)

Place an "X" in the appropriate column after each question. DK ("Don't
Know") designates uncertainty. Answer Yes or No as much as possible.

	YES	NO	DK
1. Do you think the U.S. should continue developing and testing nuclear weapons?	X		
2. Do you think that our children should be educated toward military and civil defenses against war?	X		
3. The U.S. has spent $1 billion on civil defense during the last ten years. Would you like to see some of this money being spent on health, education, and welfare instead? What percent of it? ___ %		X	
4. Do you think it is likely that preparations for war become provocations for war?		X	
5. Do you believe that President Johnson (Kennedy) really wants to disarm?	X		
6. Do you believe that Congress really wants to disarm?	X		
7. Do you believe that the Pentagon really wants to disarm?	X		
8. Would you favor efforts on our part to reduce tensions, even though this might involve some risk of having to trust Russia's peaceful intentions?		X	
9. Do you think there are any serious alternatives to our policy of deterrence, to which our leaders have not given enough attention?		X	
10. Do you believe that one person can do something to bring about world peace?		X	
11. Do you think we should fight the Communists now?	X		
12. Are you willing for the Communist Party to have access to the ballot in the U.S.?		X	
13. Are you willing for Red China to be admitted to the United Nations?		X	

14. Would you like your children to be educated in the philosophy and techniques of non-violence, such as Gandhi used in India?

_ X _

Appendix B. Some Illustrative Items from Various Scales Associated with Militarism

These are illustrative items taken from some of the scales mentioned in the introduction to this paper. The scales are listed in alphabetical order. The items are so worded that agreement with them tends to be associated with militarism.

ANTI-COMMUNISM: The Communist Party should not have access to the ballot.

ANTI-INTELLECTUALISM: Ideas are all right but it's getting the job done that counts.

ANXIETY: It is only natural for a person to be rather fearful of the future.

AUTHORITARIANISM: Obedience and respect for authority are the most important virtues children should learn.

CAPITALISM: The private enterprise system is the best way of generating and distributing wealth.

CONSERVATISM: Three meals a day will always be the best general rule.

DEMOCRATICISM (NEGATIVE): The student council should be held in check by a teacher appointed to supervise it.

DISCIPLINE: My parents tried to tell me what to do all the time.

DOGMATISM: The United States and Russia have just about nothing in common.

ETHNOCENTRISM: It would be a mistake ever to have Negroes for foremen and leaders over whites.

GOLDWATERISM: We would rather die than lose our freedom.

INTERNATIONALISM (NEGATIVE): My country should not try to improve the United Nations.

MISANTHROPY: If you don't watch out for yourself, people will take advantage of you.

NATIONALISM: My country should strive for power in the world.

NEUROTICISM: The problems I face seem to be too big for me.

OPINIONATION: It's simply incredible that anyone should believe that socialized medicine will actually help solve our health problems.

PARANOIA: I am sure I am being talked about.

PATRIOTISM: Patriotism and loyalty are the first and most important requirements of a good citizen.

PEACE RESPONSIBILITY (NEGATIVE): One person can do nothing to bring about world peace.

RELIGIOSITY: I believe in God the Father Almighty, maker of heaven and earth.

RIGIDITY: I am a methodical person in whatever I do.

SELF-REJECTION: At times I think I am no good at all.

TRADITIONAL FAMILY IDEOLOGY: Some equality in marriage may be a good thing, but by and large the husband should be the final authority in family matters.

WELFARISM (NEGATIVE): No more money should be spent on health, education, and welfare.

REFERENCES

Adorno, T. W., Frenkel-Brunswik, E., Levinson, D. J., & Sanford, R. N. *The authoritarian personality*. New York: Harper, 1950.

Alcock, N. Z. What we've learned through peace research. *United Church Observer*, December 15, 1965, 18–19, 40.

Anderson, H. H., & Anderson, G. L. Social values of teachers in Rio de Janeiro, Mexico City, and Los Angeles County, California: A comparative study of teachers and children. *Journal of Social Psychology*, 1962, 58, 207–226.

Barron, F. *Creativity and psychological health*. Princeton: Van Nostrand, 1963.

Barton, A. A survey of suburban residents on what to do about the dangers of war. *Council for Correspondence Newsletter*, March, 1963, No. 24, 3–11.

Blum, R. H. The choice of American heroes and its relationship to personality structure in an elite. *Journal of Social Psychology*, 1958, 48, 235–246.

Chesler, M., & Schmuck, R. Student reactions to the Cuban crisis and public dissent. *Public Opinion Quarterly*, 1964, 28, 467–482.

Christiansen, B. *Attitudes toward foreign affairs as a function of personality*. Oslo, Norway: Universitetsforlaget, 1959.

Comrey, A. L. Comparison of personality and attitude variables. *Educational and Psychological Measurement*, 1966, 26, 853–860.

Crow, W. J., & Noel, R. The valid use of simulation results. La Jolla: Western Behavioral Sciences Institute, June, 1965 (mimeo).

Crutchfield, R. S. Conformity and character. *American Psychologist*, 1955, 10, 191–198.

Deutsch, M. Trust, trustworthiness, and the F scale. *Journal of Abnormal and Social Psychology*, 1960, 61, 138–140.

Eckhardt, W. Ideology and personality in social attitudes. *Peace Research Reviews,* 1969a, 3 (2), whole issue.

Eckhardt, W. War in the minds of men. *War/Peace Report,* 1969b, 9 (5), 15–17.

Eckhardt, W. The factor of militarism. *Journal of Peace Research,* 1969c, 61, 123–132.

Eckhardt, W. Conservatism, East and West. *Journal of Cross-Cultural Psychology,* 1971a, 2, 109–128.

Eckhardt, W. The military-industrial personality. *Journal of Contemporary Revolutions,* 1971b, 3 (4), 74–87.

Eckhardt, W., & Alcock, N. Z. Ideology and personality in war/peace attitudes. *Journal of Social Psychology,* 1970, 81, 105 116.

Eckhardt, W., Manning, M., Morgan, C., Subotnik, L., & Tinker, L. J. Militarism in our culture today. *Journal of Human Relations,* 1967, 15, 532–537.

Eckhardt, W., & Newcombe, A. G. Militarism, personality, and other social attitudes. *Journal of Conflict Resolution,* 1969, 13, 210–219.

Ekman, P. Divergent reactions to the threat of war: Shelter and peace groups. *Science,* January 4, 1963, 139, 88–94.

Elms, A. C., & Milgram, S. Personality characteristics associated with obedience and defiance toward authoritative commands. *Journal of Experimental Research in Personality,* 1966, 1 (4).

Farris, C. D. Selected attitudes on foreign affairs as correlates of authoritarianism and political anomie. *Journal of Politics,* 1960, 22, 50–67.

Fensterwald, B., Jr. The anatomy of American isolationism and expansionism. II. *Journal of Conflict Resolution,* 1958, 2, 280–309.

Finlay, D., Iversen, C., & Raser, J. *Handbook for multi-national student survey.* La Jolla: Western Behavioral Sciences Institute, 1969.

Fraser, B. Our quiet war over peace: Politicians vs. the people. *Maclean's* January 23, 1965, 18–19, 40–41.

French, E. G., & Ernest, R. R. The relation between authoritarianism and acceptance of military ideology. *Journal of Personality,* 1955, 24, 181–191.

Gladstone, A. I. The possibility of predicting reactions to international events. *Journal of Social Issues,* 1955, 11 (1), 21–28.

Hampden-Turner, C. *Radical man: The process of psycho-social development.* Cambridge, Massachusetts: Schenkman, 1970.

Krieger, D. A developmental model of military man. *Journal of Contemporary Revolutions,* 1971, 3 (1), 68–74.

Kristjanson, L. Attitude study in Saskatchewan. In Laird, E. Organization problems in the nuclear and space age. *Saskatchewan Farmer's Union,* District 10, undated (but probably in the early 1960's).

Kuroda, Y. Correlates of the attitudes toward peace. *Background*, 1964, 8, 205–214.

Kuroda, Y. Peace-war orientation in a Japanese community. *Journal of Peace Research*, 1966, 3, 380–388.

Lane, R. E. The tense citizen and the casual patriot: Role confusion in American politics. *Journal of Politics*, 1965, 27, 735–760.

Laulicht, J., & Alcock, N. Z. The support of peace research. *Journal of Conflict Resolution*, 1966, 10, 198–208.

Lentz, T. F. Democraticness, autocraticness, and the majority point of view. *Journal of Psychology*, 1943, 16, 3–12.

Levinson, D. J. Authoritarian personality and foreign policy. *Journal of Conflict Resolution*, 1957, 1, 37–47.

MacKinnon, W. J., & Centers, R. Authoritarianism and internationalism. *Public Opinion Quarterly*, 1956, 20, 621–630.

MacKinnon, W. J., & Centers, R. Social-psychological factors in public orientation toward an out-group. *American Journal of Sociology*, 1958, 63, 415–419.

Manning, M. Militarism as prejudice. Paper presented at the Student Peace Research Workshop, Pingree Park, Colorado, August, 1964.

Mehler, D. E., Quartermain, A., Ramsay, J., & Wolins, L. Peace Union study links devout to anti-Russians. *Iowa State Daily*, March 20, 1965.

Modigliani, A. The public and the cold war. Doctor's dissertation, Harvard University, 1961.

Morris, C. W. *Varieties of human value*. Chicago: University Press, 1956.

Porter, G. Student opinions on war. Doctor's dissertation, University of Chicago, 1926.

Rokeach, M. *The open and closed mind*. New York: Basic Books, 1960.

Rokeach, M. Paradoxes of religious belief. *Trans-action*, February, 1965, 9–12.

Rokeach, M., & Eglash, A. A scale for measuring intellectual conviction. *Journal of Social Psychology*, 1956, 44, 135–141.

Rose, P. I. The public and the threat of war. *Social Problems*, 1963, 11, 62–77.

Rosenberg, M. Misanthropy and political ideology. *American Sociological Review*, 1956, 21, 690–695.

Rosenberg, M. Misanthropy and attitudes toward international affairs. *Journal of Conflict Resolution*, 1957, 1, 340–345.

Rosenberg, M. J. Images in relation to the policy process: American public opinion on cold-war issues. In Kelman, H. C. (Ed.) *International be-*

havior: A social-psychological analysis. New York: Holt, Rinehart, & Winston, 1965. Pp. 277–334.

Russell, E. W. Christianity and militarism. *Peace Research Reviews,* 1971, 4 (3).

Scott, W. A. Psychological and social correlates of international images. In Kelman, H. C. (Ed.) *International behavior: A social-psychological analysis.* New York: Holt, Rinehart, & Winston, 1965. Pp. 70–103.

Spilka, B., & Struening, E. L. A questionnaire study of personality and ethnocentrism. *Journal of Social Psychology,* 1956, 44, 65–71.

PART VIII

Suggested Additional Readings

Suggested Additional Readings

Be'eri, Eliezer, *Army Officers in Arab Politics and Society* (Albany: State University of New York Press, 1971).

Bienen, Henry (Ed.), *The Military and Modernization* (Chicago: Aldine-Atheton, Inc., 1971).

Clotfelter, James, *The Military in American Politics* (New York: Harper & Row, 1973).

Feit, Edward, *The Armed Bureaucrats: Military-Administrative Regime and Political Development* (Boston: Houghton Mifflin, 1973).

Galbraith, John Kenneth, *How to Control the Military* (New York: Signet, 1969).

Gutteridge, W., *Military Institutions and Power in the New States* (New York: Praeger, 1965).

Janowitz, Morris, *The Military in the Political Development of New Nations* (Chicago: University of Chicago Press, 1964).

Johnson, John J., *The Military and Society in Latin America* (Stanford: Stanford University Press, 1964).

Lambert, Richard D. and Adam Yarmolinsky, *Military and American Society,* American Academy of Political and Social Sciences, *Annals,* 1973.

Lefever, Ernest W., *Spear and Scepter: Army, Police, and Politics in Tropical Africa* (Washington: Brookings Institution, 1970).

Lens, Sidney, *The Military-Industrial Complex* (Philadelphia: Pilgrim Press, 1970).

Luttwak, Edward, *Coup d'Etat: A Practical Handbook* (Greenwich: Fawcett, 1969).

Just, Ward, *Military Man* (New York: Knopf, 1970).

Pye, Lucian, *Armies in the Process of Political Modernization* (Cambridge, Mass.: M.I.T. Press, 1959).

Rosem, Steven (ed.), *Testing the Theory of the Military-Industrial Complex* (Lexington: D. C. Heath, 1973).

Russett, Bruce M. and Alfred C. Stepan (eds.), *Military Force in American Society* (New York: Harper and Row, 1973).

Van Door, Jacques (ed.), *Armed Forces In Society* (The Hague: Mouton, 1968).

Welch Jr., Claude E., *Soldier and State in Africa* (Evanston: Ill.: Northwestern University Press, 1970).

Wolpin, Miles D., *Military Assistance and Counterrevolution in the Third World* (Lexington, Mass.: D. C. Heath and Co., 1973).